Seed Biology - New Advances

Edited by Ertan Yildirim, Eren Özden and Sıtkı Ermis

Published in London, United Kingdom

Seed Biology - New Advances
http://dx.doi.org/10.5772/intechopen.1000438
Edited by Ertan Yildirim, Eren Özden and Sıtkı Ermis

Contributors
Daniel Villegas, Constanza Sepúlveda, Doris Ly, Morish Obura, Jimmy Lamo, Rowshon A. Begam, Michael Deyholos, Suman Sangwan, Harshita Singh, Susheel Gulati, Lalita Singh, Archana Malik, Suryapal Singh, Peter Murithi Angaine, Alice Adongo Onyango, Jesse Owino, Khishigbuyan Turbat, Gungaanyam Galkhvv, Namjilsuren Jamiyan, Sıtkı Ermis, Eren Özden, Ertan Yıldırım

First published in London, United Kingdom, 2024 by IntechOpen
IntechOpen is the global imprint of INTECHOPEN LIMITED, registered in England and Wales, registration number: 11086078, 5 Princes Gate Court, London, SW7 2QJ, United Kingdom

British Library Cataloguing-in-Publication Data
A catalogue record for this book is available from the British Library

Additional hard and PDF copies can be obtained from orders@intechopen.com

Seed Biology - New Advances
Edited by Ertan Yildirim, Eren Özden and Sıtkı Ermis
p. cm.
Print ISBN 978-1-83769-643-7
Online ISBN 978-1-83769-642-0
eBook (PDF) ISBN 978-1-83769-644-4

For EU product safety concerns:
IN TECH d.o.o., Prolaz Marije Krucifikse Kozulić 3, 51000 Rijeka, Croatia,
info@intechopen.com or visit our website at intechopen.com.

Meet the editors

Ertan Yildirim is a full professor in the Agriculture Faculty, Department of Horticulture, Atatürk University, Turkey. His research focuses on vegetable growing, vegetable breeding, greenhouse management, seed germination and physiology, organic agriculture, and stress physiology. He was a visiting scientist at Cornell University, New York, USA. He served as Director of the Vocational Training School and Director of the Graduate School of Natural and Applied Sciences at Atatürk University, Türkiye. He is a member of national and international social and governmental organizations. He has a registered garden cress cultivar and has applied to register two bean and two pinto bean cultivars. Professor Yildirim has more than 250 publications to his credit. He has served as a reviewer for many journals. He has attended many international courses, congresses, and symposiums.

Sıtkı Ermis is Assoc. Prof. of Department of Horticulture, Agriculture Faculty at the Eskişehir Osmangazi University in Türkiye. His research is focused on teaching and researching vegetable growing, DUS testing on vegetables, seed germination and physiology, seed technology, and seed storage physiology. He was involved in many projects. He has conducted research on plant molecular genetics at the University of Cheam Crete. Before the University he worked for Registration on vegetables for a DUS Expert for 16 years and registered and conducted more than 4000 experiments with vegetable varieties. Dr. Ermiş has published more than 50 publications. He has performed review work for publications with high-impact factors. He has attended several training sessions, international courses, congresses, and symposiums.

Eren Özden completed his doctorate at Ankara University between 2012-2018. He worked as an Assistant Professor at Iğdır University, Faculty of Agriculture from 2018 to 2022. As of 2023, he has been working as an Associate Professor at Iğdır University, Faculty of Agriculture, Department of Horticulture and Kyrgyz-Turkish Manas University, Faculty of Agriculture, Department of Horticulture and Agronomy. The researcher's expertise is in seed and seedling physiology, and he focuses on seed and seedling technology, cultivation, and production of vegetable and ornamental plants. The researcher has over 60 published SCI, SCI-Expanded, International book chapters, and international publications. Among the project experiences of the researcher, he took part in a total of 15 international and national projects. The researcher received 3 international and national awards.

Contents

Preface

A seed is a living entity that contains the genetic information of the plant, ensuring reproduction and propagation. Notable progress has been made in seed development, dormancy, and germination. A fully formed seed that contains healthy embryonic tissue can produce a new plant with the right germination time and conditions.

For many higher plant species, both the starting and final material of the life cycle is the seed. Thanks to their structure, seeds can maintain their viability in temperature, humidity, and pressure environments that normal plants cannot withstand. In this way, seeds are of great importance for the survival and continuity of species and their populations.

All fully developed seeds contain an embryo, and in most plant species they have a seed coat and a nutrient reservoir that enables the embryo to develop. Seeds are stagnant during storage and will wake up and germinate when conditions of sufficient humidity and temperature are created for their growth. Each species has unique needs for the storage, germination, and development of its seeds, and these depend on light, temperature, humidity, and oxygen demands.

This book discusses seed biology, including seed morphology, physiology, metabolomics, ecology, dormancy, storage, germination, and viability. It is a useful resource for researchers as well as advanced undergraduate students and others seeking more basic knowledge on seed technology and biology.

Ertan Yildirim
Faculty of Agriculture,
Department of Horticulture,
Atatürk University,
Erzurum, Türkiye

Sıtkı Ermis
Faculty of Agriculture,
Department of Horticulture,
Osmangazi University,
Eskişehir, Türkiye

Eren Özden
Faculty of Agriculture,
Department of Horticulture,
Iğdır University,
Iğdır, Türkiye

Faculty of Agriculture,
Department of Horticulture and Agronomy,
Kyrgyz-Turkish Manas University,
Bishkek, Kyrgyzstan

Chapter 1

Use of Low-dose Gamma Radiation to Promote the Germination and Early Development in Seeds

Daniel Villegas, Constanza Sepúlveda and Doris Ly

Abstract

The study of the effect of low doses of ionizing radiation on the germination and initial growth of different seeds is a recent area of research, with gamma rays and X-rays receiving the most attention. The use of this type of energy can generate an increase in germination percentages, an increase in germination speed, and changes in the length and area of roots and shoots, which will depend both on intrinsic factors of the nature of the energy (dose, dose rate, energy, etc.) as well as aspects of the irradiated seeds (water content, sensitivity, etc.). In addition to morphological effects, radio-stimulation due to low doses of ionizing radiation (a phenomenon also described as radio-hormesis) generates changes at physiological, biochemical, metabolic, and molecular levels. Despite the evidence that has been accumulating, it is still necessary to deepen the knowledge about these phenomena in order to establish the use of ionizing radiation with the aim of using radio-stimulation as a real impact tool in the agroforestry sector.

Keywords: radio-stimulation, hormesis, ionizing radiation, germination, plant development

1. Introduction

The mutagenic effect of ionizing energy has been widely studied, being the most commonly used physical agent to irradiate seeds (and other plant materials) in order to generate heritable mutations in plant breeding programs [1]. For this purpose, it is necessary to define the highest possible dose to induce high frequency mutations and, at the same time, the least negative effects dose, a concept named Lethal Dose 50 (LD50) [2]. However, in recent years, the non-mutagenic effect related to low doses (below LD50) of ionizing radiation (IR) is an area that has attracted special attention, a phenomenon known as radio-stimulation or radio-hormesis (from the Greek "hormaein", which means to stimulate) [3]. The vast majority of plant foods are produced by crops that are propagated by seeds, but the germination process is highly vulnerable to external conditions, which can cause delayed and/or uneven germination or even weak seedlings, which will inevitably end up affecting the yield and production quality [4]. In search of alternatives to reduce the difficulties associated with seed quality, low doses of IR have begun to be applied to speed up the germination processes, increase seedling quality, and to improve tolerance to biotic and abiotic stressors [5].

Within the different types of IR (X-rays; alpha and beta particles; neutrons), the majority of research has been done using gamma radiation due to the safety and ease of operation of equipment such as Gammacell irradiators.

Gamma rays can engage with cell components like atoms or molecules. This interaction affects the structure and biochemical processes and, in this way, modifies the overall plant behavior. Seed germination, seedling growth, secondary metabolite synthesis, and biotic/abiotic stress responses are some of the main processes that are modified in response to gamma ray exposure [5, 6].

Considering the large amount of information accumulated to date on this topic, the purpose of this chapter is to summarize the most recent and more relevant results on the radio-stimulation of seeds.

2. Physical nature of IR and its interaction with plant tissues

To understand the stimulating effect of IR, it is necessary to recall basic aspects of atomic structure. All atoms are made up of protons and neutrons (forming the atomic nucleus) and electrons that "orbit" around the nucleus. When the nuclear forces that keep the protons and neutrons together are strong enough to overcome the electric repulsion between particles of the same charge (protons), the atom will remain stable. Conversely, an atom will be unstable when the number of neutrons exceeds a limit ($z = 82$), causing the excess energy to be released in the form of radioactivity to maintain the integrity of the nucleus [7]. The radioactivity emitted from the nucleus can be in the form of alpha particles, beta particles, gamma radiation, or a combination of the three. Alpha particles are composed of 2 protons and 2 neutrons (structurally equivalent to a positively charged Helium nucleus), while beta particles are electrons that come from the nucleus (the product of the transformation of a neutron into a proton and an electron). The formed electron is released from the nucleus as a result of the decay [8]. Due to the release of alpha and beta particles, a rearrangement occurs inside the nucleus, which in turn results in the release of electromagnetic energy or gamma radiation [9].

2.1 Interaction of IR with matter

Both alpha and beta particles, as well as gamma radiation, have the ability to interact with matter and deposit enough energy to "knock out" an electron and thereby ionize the matter [10]. This ability to alter the atomic structure (ionization) is what differentiates the effect of alpha particles, beta particles, and gamma radiation from the effect of UV light and other electromagnetic radiations (visible light, infrared, microwaves, etc.), since the latter are capable of modifying the behavior of matter but do not have enough energy to alter its structure [7]. To understand the differences between alpha particles, beta particles, and gamma radiation, it is necessary to define two main components that largely determine the ability of each of these to interact with matter and, consequently, modify its behavior: Penetrability and Linear Energy Transfer (LET). Penetrability refers to the ability of radiation to pass through matter, which is directly related to the mass and electric charge of the radiation. Alpha particles have a mass (2 protons +2 neutrons) and an electric charge of +2, which means that they are highly reactive particles and therefore have very low penetrability (they are blocked by a paper sheet) [8]. Beta particles, with a fraction of the mass of a proton and an electric charge of −1, penetrate material to a greater depth than alpha particles [9] and can be blocked by aluminum foil. Gamma radiation, due to its electromagnetic nature, does not have mass, and therefore it can penetrate deeply into matter. To block gamma radiation, dense materials such as concrete or lead are required [11]. Linear Energy Transfer (LET) refers to the energy dissipated per unit length along the tracks of the ionizing particles [12]. Gamma rays have low linear energy transfer (LET), while alpha particles have high linear energy transfer (LET) [13]. The radiation source, along with penetrability and linear energy transfer (LET), determines the amount of energy absorbed by matter or absorbed dose, which is expressed in Gy (gray), with 1 Gy equivalent to 1 Joule of energy absorbed per kilogram [14].

At the level of plant tissues or cells, the interaction of IR can occur in two ways. When the energy of the radiation is deposited directly onto macromolecules (DNA, membranes, etc.) [15], it generates "damage" that, depending on the dose, can become lethal [16, 17]. However, the main effect of IR on plant cells occurs indirectly and it is mediated by the ionization of the water molecule (radiolysis), which results in the production of Reactive Oxygen Species (ROS) [18, 19]. Most of the responses described below, related to the stimulating effects of low doses of IR applied to seeds, are ultimately mediated by the accumulation of ROS.

For more detailed information on the physical, physicochemical, and chemical processes triggered by the interaction of IR with the water molecule, please refer to [20, 21].

3. The effects of low doses of IR on germination, plant growth, and development

Before presenting reports on the effect of treating seeds with low doses of IR, it is important to emphasize that the range of what is called low doses of IR is species-specific and depends not only on the particular radio-sensitivity of each species but also on the type of radiation (alpha particles, beta particles, gamma rays, X-rays, etc.), the rate of dose (acute or chronic irradiation), the pre-treatment of the material to be irradiated (moisture content), and the ontogenic state of the irradiated material [22]. In this way, the stimulating dose of IR found in the literature can vary from less than 1 Gy to doses of more than 1 KGy [23, 24], which highlights the difficulty of establishing cross-sectional ranges of stimulating thresholds as these must be defined on a case-by-case basis. In view of the above, the data and results presented below aim only to compile the growing and recent information on radiation stimulation by IR in seeds and visualize the differential effect (stimulation or inhibition) that different IR treatments have on different plant species. A simplified scheme summarizing the overall seed radio-stimulation response to low doses of IR is presented in **Figure 1**.

3.1 IR and effect on germination

The effects of low doses of IR on plants have been studied for several decades. These studies have been conducted on different species and varieties, applying IR to different structures (seeds or vegetative structures), under different systems (*in vivo* or *in vitro*), on different ploidy levels and/or ages of the organ or tissues and using different dose and dose/rate combinations [25]. All of these parameters make it difficult to define a standard protocol to apply to a given species or situation.

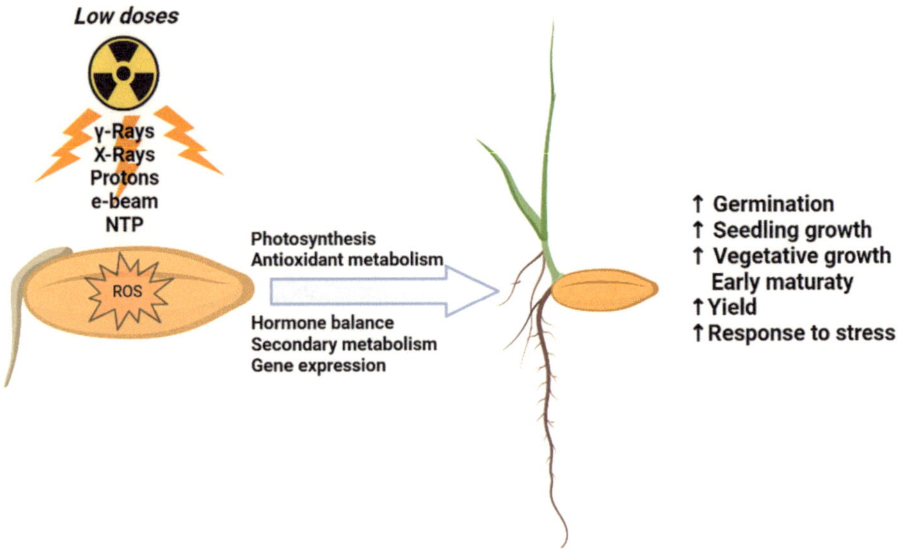

Figure 1.
Overall radio-stimulation response of seed treated with low doses of ionizing radiation. Created with BioRender.com.

Despite the above, in literature, some authors have made efforts to define low doses like the one that falls between 5 and 20 Gy for seeds and 1 to 5 Gy for vegetative material [26]. These authors mention only stimulatory effects of low doses, on seeds of several crops, like *Capsicum annuum*, *Arabidopsis thaliana*, *Phaseolus vulgaris*, *Cajanus cajan*, *Triticum sp*. and in vegetative stages of *Arabidopsis thaliana*. Another study on tomato seeds [27] defines 150 Gy as a low dose that stimulates parameters like germination, fruit number, and total production up to 86%.

Nevertheless, it has been observed that doses ranging from 5 to 800 Gy of gamma radiation have had stimulatory effects on growth of dry seeds [28]. This wide range of possibly stimulating doses is related to seed radio-sensitivity, which in terms depends on genetic characteristics and on seed moisture content. **Table 1** shows the latest works on the stimulation effect caused by IR.

It has been reported repeatedly that low dose rate and/or low total dose gamma irradiation impact germination yield and seedling performance, acting like an actual priming treatment [14]. Due to this well-documented effect, efforts have been made to investigate the molecular mechanism activated in seeds as a response to this physical treatment. Doses lower than 100 Gy of gamma rays positively stimulated the germination index, seedling growth, primary root length, and fresh weight on *A. thaliana* seeds. In that work, 50 Gy was the dose that showed the maximal positive effect on all growth parameters [43].

Seed germination, vigor, and seedling growth in wild oat (*Avena fatua L.*) [44], garden cress (*Lepidium sativum L.*) [45], deadly nightshade (*Atropa belladonna L.*) [46], okra (*Abelmoschus esculentus L.* Monch.) [47], and rocket (*Eruca vesicaria L. subsp. sativa*) [48] have been stimulated by low-dose gamma rays. All these works provided cumulative evidence that small doses of γ-rays result in beneficial action in physically treated seeds. These works point to several mechanisms as responsible for the effect of low-dose gamma rays on seeds and early growth. Among the triggering factors of the response are the ROS produced, which act as signaling molecules to

Plant species	Stimulatory dose (Gy)	Reference
Sugarcane	4.7–5.7	[29]
Datura innoxia	5	[30]
Eucalyptus nitens	10	[31]
Hordeum vulgare	4–20	[32–34]
Cucumis sativus	50	[35]
Abelmoschus esculentus (seeds, seedlings)	50	[35]
Chenopodium quinoa	50	[36]
Vicia faba	<100	[37]
Lathyrus chrysanthus	50–150	[38]
Vigna unguiculata	100	[39]
Physalis peruviana	125–200	[40]
Abelmoschus esculentus	400	[41]
Sophora davidii	800	[42]

Table 1.
Low-dose IR on some seeds.

respond under stress conditions; increased enzymatic activity; nucleic acids and protein synthesis in treated seeds. These changes in metabolism could explain the boost in germination, break of dormancy, and plant development. Conversely, exposing seed to high doses of gamma rays has been demonstrated to alter protein synthesis, hormonal equilibrium, enzyme activity, and leaf gas, and water exchange [49].

3.2 IR and its effect on growth

There are reports showing that low gamma irradiation doses led to positive effects on growth and plant yield in tomato hybrids [50]. Nevertheless, El-Sayed et al. [51] reported that 12 krad (equivalent to 120 Gy) gamma rays increased plant height, yield, chlorophyll a and b and carotenoids in tomato hybrids. Studies on Jerusalem artichoke (*Helianthus tuberosus*) show a stimulatory effect at 5 Gy dose on plant height, number of branches, fresh and dry shoot weight [52].

The effect of gamma rays on plant growth and development is explained by cytological, genetical, biochemical, physiological, and morphogenetic changes in cells and tissues [53]. These changes are commonly reported as more vigorous vegetative growth [54], early maturity [55], and higher yield [56].

While a definitive explanation regarding positive impacts of low-dose gamma radiation remains elusive, researchers suggest a theory that these irradiation levels stimulate growth by altering the hormonal signaling network within plant cells. Besides, it is proposed that increased cells' antioxidative capacity could allow better performance over stressful fluctuations in light intensity and temperature conditions [57]. Conversely, the growth suppression due to high doses of irradiation has been linked to cell cycle alterations in the G2/M phase, as well as several impairments across the entire genome [58].

Recent studies have shown that not only the dose is important but also the dose rate. Even when the final dose was the same, long-term exposure to gamma rays produced more free radicals than shorter exposure. When exposed for short periods, wheat shoot and root lengths showed minor decreases compared to control samples, however, longer periods resulted in substantial growth reduction. The expression of genes associated with antioxidants and DNA repair showed a reduction in response to long-term gamma ray exposure [59].

3.3 IR and seed priming

Seed priming corresponds to the induction of a particular physiological state through the application of treatments (physical, chemical, thermal, etc.) prior to germination, which allows the plant to better respond to the subsequent presence of abiotic and biotic stresses [60]. In recent years, priming (particularly at the seed level) has emerged as a strategy for stress management without significantly affecting plant development [61]. Many of the advances in understanding and elucidating the stimulatory effect of low doses of ionizing radiation have their origin in the study of priming through physical stimuli [14]. A study on the effect of low doses of gamma radiation and the induction of tolerance to stress caused by Cadmium and Lead in *Arabidopsis* seeds [43], reported that doses up to 100 Gy induced better germination rates and initial growth. It also demonstrated that doses of 50 Gy induced a better response of tolerance to stress caused by these metals, such as a decrease in the presence of H_2O_2, higher activity of antioxidant enzymes, and greater accumulation of proline compared to non-irradiated seeds. Similar results were observed when irradiating *Hordeum vulgare* seeds with doses up to 300 Gy, where those seedlings derived from seeds irradiated with 50 Gy improved their tolerance to the presence of heavy metals (lower contents of H_2O_2 and improvement in the ultrastructure of chloroplasts) [62]. Doses of 50 Gy applied to *Arabidopsis* seeds stimulated the tolerance of seedlings to thermal stress (improved growth rates, reduced ROS levels, higher antioxidant enzyme activity, etc.) [63], while exposure to 100 Gy applied to *Vicia sativa* L. seedlings (alone or in combination with salt and drought stress) generated significant increases in dry matter accumulation, higher antioxidant enzyme activity (CAT, SOD and APX), higher proline contents, and decreases in relative water content [6]. Doses between 500 and 1000 Gy decreased the incidence of fungal diseases in *Pennisetum glaucum* grains, and despite the high doses applied, no effects on the germination percentage of these grains were observed [64].

3.4 IR and metabolic effects

Due to its penetrability and linear energy transfer, as gamma radiation passes through different plant structures and tissues, it generates a broad spectrum of modifications affecting biochemical, physiological, and molecular processes. As the effect of IR is mainly mediated by the increase in ROS, a significant part of the literature focuses on describing the processes that are directly modified by these molecules, such as processes associated with photosynthesis or processes associated with the plant antioxidant system [26].

3.4.1 Effect on photosynthesis

IR affects various components of the photosynthetic apparatus, such as the content of pigments responsible for the absorption of visible radiation; enzymes responsible for CO2 reduction; thylakoids structure, etc. [26]. Regarding the chlorophyll content, various studies show contradictory trends or dynamics. Studies in

Arabidopsis thaliana show that chlorophyll concentration remains stable up to doses of approximately 60 Gy [65], while similar doses (50 Gy) generated a significant increase in the total chlorophyll content in cowpea [66].

On the other hand, a recent study on soybean seeds and seedlings reported a positive relationship between the content of chlorophyll a, chlorophyll b, and carotenoids at a dose of 12 Gy when compared to non-irradiated seeds [67]. Another study suggests that, in general, the effect of IR on photosynthetic pigments follows a pattern of mild stimulation at low IR doses, while as the doses increase, the concentrations may initially increase but then drop in the long term [26]. The activity of the Rubisco enzyme, a central component of the CO2 fixation process, shows stimulation at low doses (5 and 25 Gy) in wheat seedlings [68], while higher doses decrease the specific activity of this enzyme in *Arachis hypogaea L* [69].

The radio-sensitivity studies of this enzyme are particularly complex since the different subunits that form this enzyme are encoded by both nuclear and plastidial genetic material [70]. However, despite the stimulatory effect of low doses of radiation on Rubisco activity, the effect of radiation on the rate of CO2 assimilation appears to be negative even at doses as low as 0.12 Gy [71] and 1.2 Gy [72], while higher doses have resulted in prolonged inhibition of the CO2 assimilation rate [68].

3.4.2 Antioxidant metabolism

The increase of ROS after exposure to IR has been thoroughly discussed. Seeds and plants subjected to IR show high levels of ROS that remain elevated for times ranging from several hours to several days post-irradiation. Since ROS are short-lived compounds, the increase of these components observed in plant tissues days or hours post-irradiation would not be the product of the direct process of radiolysis of water but rather the result of an imbalance between the processes of generation and use of these compounds [26]. In this way, it is expected that IR affects the content and/or activity of various enzymes involved in these processes. For example, irradiation with doses between 25 and 200 Gy stimulates the activity of ascorbate peroxidase and glutathione reductase enzymes in rice seeds [73], while peroxidase activity increased in orange seeds (doses of 10–50 Gy) [74] and in bean seeds (150 and 200 Gy) [75]. Irradiation of red pepper seeds with doses of 2 to 16 Gy resulted in an increase in superoxide dismutase activity but, at the same time, led to a decrease in glutathione reductase activity [57].

Several studies have established that ROS production mediated by IR is dose-dependent and follows a trend close to linearity [73, 76–78]. However, the existence of multiple pathways for ROS production and utilization (in response to biotic and abiotic stimuli) makes it very complex to identify or describe a specific pathway of response to IR mediated by ROS in plants [18]. Besides, these pathways are being influenced or affected differentially depending on the radiation dose [79].

3.4.3 Effect on secondary metabolism

One of the most widely studied processes is the stimulating effect of IR on secondary metabolism, and more specifically on metabolites with antioxidant capacity. It is postulated that the increase in antioxidant compounds would be triggered by the high concentration of ROS resulting from the radiolysis of water through two different pathways [80]. On one hand, the increase in ROS would stimulate the expression of genes that encode for key enzymes in the biosynthetic pathways of these compounds, as demonstrated by the results in *Rosmarinus officinalis* [81] and *Arabidopsis thaliana* [82]. On the other hand, studies on different medicinal plants show that the presence of ROS would directly stimulate the activity of these enzymes and, even more interestingly, demonstrate that the increase in activity of enzymes such as phenylalanine ammonia-lyase (PAL) and chalcone synthase (CHS) is dose-dependent [83]. *Solanum melongena* seeds treated with 50 Gy gamma rays showed increased growth that led to increased levels of flavonoid and tannin contents in pulp, peel, and whole fruits [84].

3.4.4 Molecular response to low doses of IR

As detailed in the previous sections, the radio-stimulative effect of gamma radiation on seeds encompasses a wide range of processes and responses, ranging from promoting germination to modifications in plant fruiting. This diversity of effects has led to the absence, to date, of a single or consensus mechanism that molecularly modulates these responses [28]. Apart from the damage (repairable

Gen name	Plant sp.	Dose	Function	Reference
Apetala 1 (AP1)	*Arabidospis thaliana*	3 cGy and 17 cGy (chronic exposure); 15 Gy (acute exposure)	Promote floral meristem identity	[86]
Ascobate peroxide (APX), Catalase (CAT), and Glutathione Reductase (GR)	*Oryza sativa*	25, 50, 100, and 200 Gy	Antioxidant defense	[73]
Heat Shock Protein 90 (HSP90)	*Arabidopsis thaliana*	100, 250, 500, and 750 Gy	Signal transduction, cell cycle, DNA repair, and stress response	[87]
Lipid Transfer Protein (EARLI 1)	*Hordeum vulgare*	15, 20, and 25 Gy	Germinability and early seedling development	[34]
Proliferating Cell Nuclear Antigen (PCNA)	*Hordeum vulgare*	1.6, 2.6, and 4.2 μSv/h	Cell cycle regulation	[77]
Superoxide Dismutase (SOD) and Guaiacol Peroxidase (GPOX)	*Hordeum vulgare*	2, 4, 6, 8, 10, 13, 16, 20, 25, and 50 Gy	Antioxidant defense	[33]
Suppressor of Gamma Response 1 (SOG1)	*Arabidopsis thaliana*	1 to 540 mGy/h	Master regulator of DNA damage response	[88]
Terpene Synthase 30 (TPS30)	*Oryza sativa*	100, 200, 300, 400, and 500 Gy	Secondary metabolism	[89]
X-Ray Repair Cross-Complementing Protein (XRCC)	*Oryza sativa*	100, 300, and 500 Gy	DNA damage response	[59]

Table 2.
Seed gene expression modified by IR.

or unrepairable) that IR causes at the DNA level (associated with higher doses, which finally led to mutations), IR also causes modifications in the regulation of the genome in processes related to oxidative stress, signal transduction, transcription factors, hormone response, metabolism transport, energy, development and morphogenesis, and cell cycle [26, 82, 85]. Due to the extensive nature of a section that would describe the background information available in the literature regarding this topic (including plant species, radiation dose, acute vs. chronic exposure, etc.), **Table 2** presents some relevant reports of genes that modify their expression in response to low doses of IR in seeds. For further information, the reader is encouraged to review [21, 26, 85, 90].

4. Effect on germination and growth of other physical agents

Much of the research on the effects of physical agents on seed growth and development is carried out in self-shielded equipment with 60Co as a gamma radiation emitting source (i.e., Gammacell 220R). However, there are other types of physical agents not necessarily derived from nuclear reactions that have also been used to study their possible radio-stimulating effect.

4.1 X-rays

X-rays correspond to a non-nuclear IR of electromagnetic nature (similar to gamma rays) that has also been used to modify the behavior and development of various plant species. Regarding their use as a radio-stimulant agent, shorter germination times have been reported in *Hibiscus* [91], increased vigor of coffee seedlings [92], increased leaf area in *Phaseolus* plants [93], and increased leaf and plant size in Tomato [25] in response to doses ranging from 0.1 Gy to doses greater than 100 Gy.

4.2 Protons

It corresponds to corpuscular radiation (hydrogen nuclei) typical of extraterrestrial environments (part of solar particles). The study of the effect of protons on the growth and development of plant species is crucial to evaluating potential species to be used in space missions or settlements [94]. In this regard, there are reports of higher germination rates and increases in chlorophyll and ascorbate peroxidase (APX) contents in soybean plants irradiated with protons [95], as well as stimulation of seedling growth and greater plant height [96] and root elongation [97] in rice seeds irradiated with low doses of this type of ionizing energy.

4.3 Electron beam

Low-energy electron beam (LEEB) beta radiation consists of a beam of accelerated electrons. This form of ionizing radiation operates within a range spanning from a few up to around 300 kGy. The accelerated electrons display enough energy to remove electrons from atoms or molecules producing ions [98]. LEEB application to lentil seeds accelerated seed germination, defined by the percentage of hypocotyl and leaf emergence at 3 days [99]. There are also reports that low doses of this type of radiation, applied to barley seeds, induced higher germination rates [100] while it improved the height and weight in wheat plants [101].

4.4 Non-thermal plasma (NTP)

By increasing the internal energy of a material, it will go from solid to liquid to gas and finally to an ionized gas state (where electrons separate from the elements), giving rise to the fourth state of matter or "plasma", which has unique properties [102]. Depending on the conditions (working pressure, type of energy, amount of energy) required for plasma formation, it will have different properties [103], with non-thermal plasma (NTP) being the most studied for inducing changes at the seed or seedling level because the low temperatures generated do not alter the behavior of the material subjected to such plasma. Since the effect depends not only on the dose or energy imparted or absorbed by the irradiated material but also on the conditions under which the NTP state is induced or reached, it is complex to summarize the results obtained (and beyond the scope of this article). As an example, a report from 2005 studied the effect on the growth and development of tomato plants obtained from seeds irradiated with NTP [104], reporting higher seedling emergence, greater antioxidant enzyme activity, and a higher number of fruits and fruit biomass per plant. Stimulating effects of different doses and configurations of NTP have also been reported in cereals [105], legumes [106, 107], oilseeds [108], vegetable crops [109, 110], among others.

An alternative to the direct exposure of plant material (usually seeds) to NTP is the exposure of water to this type of energy, resulting in what is known as plasma-activated water (PAW) or plasma-treated water (PTW), which induces several chemical and physical modifications resulting in different biological effects on plant material [111]. The response to PAW involves change in redox potential, conductivity, pH and ROS and reactive nitrogen species content [112]. By using this technology, *Vigna radiata* seeds exposed to different treatments with PAW (different time exposition of water to NTP) resulted in the maximization of parameters such as germination rate and growth parameters, as well as an increased content of flavonoids and total phenols in seedlings [113]. Wheat seeds treated with PAW resulted in better germination, faster seedling growth, higher photosynthetic pigments in leaves, and soluble protein content in roots [114].

5. Conclusions

According to statistics, 90% of edible crops are cultivated from seeds [5]. Low germination and poor seedling growth often result in huge crop losses and therefore, developing strategies aimed at improving processes related to seed germination and crop establishment is a primary way to ensure food security [4, 115]. The necessity to develop strategies applied to seeds that aim to improve these processes while being environmentally friendly has driven studies on the use of low doses of IR (mainly gamma radiation) as a stimulating agent, a phenomenon called radio-stimulation or radio-hormesis. In the last two decades, increasing evidence has accumulated of the radio-stimulatory effect of gamma radiation (a safe, non-polluting, and sustainable form of ionizing radiation) on seed germination and associated processes.

The range or dose limit in which the stimulatory effect of gamma radiation is observed depends on intrinsic factors of the applied energy (LET, dose, dose rate) as well as the irradiated material (species; tissue; state of development, etc.). This makes it difficult to use transversal concepts regarding dose limits, that can be called "low doses" and rather, the stimulatory effect must be studied and defined on a case-by-case basis.

In the same way, the physiological processes and metabolic responses that are modified as a result of the application of stimulating doses of IR are also diverse and include changes in hormonal balance, activation of antioxidant protection systems, modifications of parameters associated with photosynthesis, stimulation of secondary metabolites, etc. However, despite decades of research, the precise mechanism of beneficial plant response to IR remains elusive [28], although there seems to be a consensus that all responses are associated (directly or indirectly) with changes in the content of ROS species.

Recently, the use of new techniques such as ion beam and especially the use of NTP have emerged as cost-effective and non-polluting alternatives capable of stimulating germination and modifying the development of some plant species [4].

Despite the advances and studies on the stimulating effect of gamma radiation on seed germination and eventually on the growth and development of plants, the use of this technique is still mostly associated with scientific studies and little progress has been made on its operational use. Therefore, it is necessary to continue researching to elucidate the response and molecular pathways that modulate the interaction between the ionizing radiation and plant development with the aim of making use of radio-stimulation as a real impact tool in the agroforestry sector.

Acknowledgements

This work has been supported by the Grant FONDECYT-PAI 7818I20007 granted by Research & Development National Agency (ANID).

Author details

Daniel Villegas[1*], Constanza Sepúlveda[2] and Doris Ly[1]

1 Chilean Nuclear Energy Commission, Santiago, Chile

2 Adolfo Ibáñez University, Santiago, Chile

*Address all correspondence to: daniel.villegas@cchen.cl

References

[1] Shoba Sivasankar S, Noel Ellis N, Jankuloski L, Ingelbrecht I, editors. Mutation Breeding, Genetic Diversity and Crop Adaptation to Climate Change. Wallingford, England: CABI; 2021

[2] Kodym A, Afza R, Forster BP, Ukai Y, Nakagawa H, Mba C. Methodology for physical and chemical mutagenic treatments. In: Plant Mutation Breeding and Biotechnology. UK: CABI; 2012. pp. 169-180

[3] Miller MW, Miller WM. Radiation hormesis in plants. Health Physics. 1987;**52**(5):607-616. DOI: 10.1097/00004032-198705000-00012

[4] Bera K, Dutta P, Sadhukhan S. Seed priming with non-ionizing physical agents: plant responses and underlying physiological mechanisms. Plant Cell Reports. 2022;**41**(1):53-73. DOI: 10.1007/s00299-021-02798-y

[5] Wang J, Zhang Y, Zhou L, Yang F, Li J, Du Y, et al. Ionizing radiation: Effective physical agents for economic crop seed priming and the underlying physiological mechanisms. International Journal of Molecular Sciences. 2022;**23**(23):15212. DOI: 10.3390/ijms232315212

[6] Beyaz R. Impact of gamma irradiation pretreatment on the growth of common vetch (*Vicia sativa* L.) seedlings grown under salt and drought stress. International Journal of Radiation Biology. 2020;**96**(2):257-2669. DOI: 10.1080/09553002.2020.1688885

[7] Mba C, Afza R, Shu QY. Mutagenic radiations: X-rays, ionizing particles and ultraviolet. In: Shu QY, Forster BP, Nakagawa H, editors. In Plant Mutation Breeding and Biotechnology. Wallingford, England: CABI Publishing; 2012. pp. 83-90

[8] L'Annunziata MF. Radioactivity: Introduction and History, from the Quantum to Quarks. 2nd ed. London, England: Elsevier Science; 2016

[9] IAEA. Radiation Biology: A Handbook for Teachers and Students. IAEA. [cited 2023 Jun 5];1. 2019. Available from: https://www.iaea.org/publications/8219/radiation-biology-a-handbook-for-teachers-and-students

[10] Sankaranarayanan K. Estimation of the hereditary risks of exposure to ionizing radiation: History, current status, and emerging perspectives. Health Physics. 2001;**80**(4):363-369. DOI: 10.1097/00004032-200104000-00013

[11] Barrachina Gomez M, Cerrolaza Asenjo JA, Garcia Alonso JM, Iranzo Martin JE, Lopez Perez B, Minguez Perres E, et al. 222 Questions about the Energy. Spain: Forum Atómico Espanol; 1993

[12] Spencer-Lopes MM, Forster BP, Jankuloski L. Manual on Mutation Breeding. Rome, Italy: FAO; 2018

[13] De Micco V, Arena C, Pignalosa D, Durante M. Effects of sparsely and densely ionizing radiation on plants. Radiation and Environmental Biophysics. 2011;**50**(1):1-19. DOI: 10.1007/s00411-010-0343-8

[14] Araújo S, Paparella S, Dondi D, Bentivoglio A, Carbonera D, Balestrazzi A. Physical methods for seed invigoration: Advantages and challenges in seed technology. Frontier in Plant Science. 2016;7:646. DOI: 10.3389/fpls.2016.00646

[15] Vanhoudt N, Cuypers A, Vangronsveld J, Horemans N, Wannijn J,

Van Hees M, et al. Study of biological effects and oxidative stress related responses in gamma irradiated *Arabidopsis thaliana* plants. Radioprotection. 2011;**46**(6):S401-S407. DOI: 10.1051/radiopro/20116510s

[16] Culligan KM, Robertson CE, Foreman J, Doerner P, Britt AB. ATR and ATM play both distinct and additive roles in response to ionizing radiation. The Plant Journal. 2006;**48**(6):947-961. DOI: 10.1111/j.1365-313X.2006.02931.x

[17] Esnault M-A, Legue F, Chenal C. Ionizing radiation: Advances in plant response. Environmental and Experimental Botany. 2010;**68**(3):231-237. DOI: 10.1016/j.envexpbot.2010.01.007

[18] Gudkov SV, Shilyagina NY, Vodeneev VA, Zvyagin AV. Targeted radionuclide therapy of human tumors. International Journal of Molecular Sciences. 2015;**17**(1):33. DOI: 10.3390/ijms17010033

[19] Bruggeman PJ, Kushner MJ, Locke BR, Gardeniers JGE, Graham WG, Graves DB, et al. Plasma–liquid interactions: a review and roadmap. Plasma Sources Science and Technology. 2016;**25**(5):053002. DOI: 10.1088/0963-0252/25/5/053002

[20] Sartorio C, Angiolini M, Flammini D, Pietropaolo A, Agostini P, Alberghi C, et al. Preliminary assessment of radiolysis for the cooling water system in the rotating target of SORGENTINA-RF. Environments. 2022;**9**(8):106. DOI: 10.3390/environments9080106

[21] Caplin N, Willey N. Ionizing radiation, higher plants, and radioprotection: From acute high doses to chronic low doses. Frontier in Plant Science. 2018;**2018**:9. DOI: 10.3389/fpls.2018.00847

[22] Volkova PY, Duarte GT, Kazakova EA, Makarenko ES, Bitarishvili SV, Bondarenko VS, et al. Radiosensitivity of herbaceous plants to chronic radiation exposure: Field study in the Chernobyl exclusion zone. Science Total Environment. 2021;**777**(146206):146206. DOI: 10.1016/j.scitotenv.2021.146206

[23] Jan S, Parween T, Siddiqi TO, Mahmooduzzafar. Effect of gamma radiation on morphological, biochemical, and physiological aspects of plants and plant products. Environmental Reviews. 2012;**20**(1):17-39. DOI: 10.1139/a11-021

[24] Majeed A, Muhammad Z, Ullah R, Ali H. Gamma irradiation I: effect on germination and general growth characteristics of plants–a review. Pakistan Journal of Botany. 2018;**50**(6):2449-2453

[25] De Micco V, Paradiso R, Aronne G, De Pascale S, Quarto M, Arena C. Leaf anatomy and photochemical behaviour of *Solanum lycopersicum* L. plants from seeds irradiated with low-LET ionising radiation. Scientific World Journal. 2014;**2014**:428141. DOI: 10.1155/2014/428141

[26] Gudkov SV, Grinberg MA, Sukhov V, Vodeneev V. Effect of ionizing radiation on physiological and molecular processes in plants. Journal of Environmental Radioactivity. 2019;**202**:8-24. DOI: 10.1016/j.jenvrad.2019.02.001

[27] Prabhat VK. Effects of gamma radiation on tomato seeds. Ijsdr.org. [cited 2023 Jun 5]. 2023. Available from: https://www.ijsdr.org/papers/IJSDR2010013.pdf

[28] Volkova PY, Bondarenko EV, Kazakova EA. Radiation hormesis in plants. Current Opinion in Toxicology.

2022;**30**:100334. DOI: 10.1016/j.cotox.2022.02.007

[29] Side THR, Abdurrakhman A, Djumali D, Herwati A, Yulaikah S, Supriyono S. Developing determination of gamma irradiation dose to increase sugarcane growth and yield. E3S Web Conference. 2023;**373**:03013. DOI: 10.1051/e3sconf/202337303013

[30] Aref IM, Khan PR, Al Sahli AA, Husen A, Ansari MKA, et al. Response of *Datura innoxia* Linn. To gamma rays and its impact on plant growth and productivity. Proceedings of the National Academy Science India Sect B Biol Sci. 2016;**86**(3):623-629. DOI: 10.1007/s40011-014-0485-6

[31] Gutiérrez B, Koch L, Villegas D, Gonzalez J, Ly D, Molina M, et al. Análisis de Germinación de Semillas de Eucalyptus nitens Tratadas con Radiación Gamma: Indicios de Efecto Hormético. C&I Forestal. 2021;**27**(3):7-16. DOI: 10.52904/0718-4646.2021.554

[32] Volkova PY, Clement G, Makarenko ES, Kazakova EA, Bitarishvili SV, Lychenkova MA. Metabolic profiling of γ-irradiated barley plants identifies reallocation of nitrogen metabolism and metabolic stress response. Dose-Response. 2020;**18**(1):1559325820914186. DOI: 10.1177/1559325820914186

[33] Geraskin S, Churyukin R, Volkova P. Radiation exposure of barley seeds can modify the early stages of plants' development. Journal of Environmental Radioactivity. 2017;**177**:71-83. DOI: 10.1016/j.jenvrad.2017.06.008

[34] Volkova PY, Duarte GT, Soubigou-Taconnat L, Kazakova EA, Pateyron S, Bondarenko VS, et al. Early response of barley embryos to low- and high-dose gamma irradiation of seeds triggers changes in the transcriptional profile and an increase in hydrogen peroxide content in seedlings. Journal of Agronomy and Crop Science. 2020;**206**(2):277-295. DOI: 10.1111/jac.12381

[35] Jaipo N, Kosiwikul M, Panpuang N, Prakrajang K. Low dose gamma radiation effects on seed germination and seedling growth of cucumber and okra. Journal of Physics Conference Series. 2019;**1380**(1):012106. DOI: 10.1088/1742-6596/1380/1/012106

[36] Song KE, Lee SH, Jung JG, Choi JE, Jun W, Chung J-W, et al. Hormesis effects of gamma radiation on growth of quinoa (*Chenopodium quinoa*). International Journal of Radiational Biology. 2021;**97**(7):906-915. DOI: 10.1080/09553002.2021.1919783

[37] El-Gazzar N, Mekki L, Heneidak S. ISSR markers associated with effects of gamma irradiation on growth and seed yield of M2 plants of Faba bean (*Vicia Faba* L). AJSRP. 2016;**2**(2):75-89. DOI: 10.12816/0025266

[38] Beyaz R, Kahramanogullari CT, Yildiz C, Darcin ES, Yildiz M. The effect of gamma radiation on seed germination and seedling growth of Lathyrus chrysanthus Boiss. under in vitro conditions. Journal of Environmental Radioactivity. 2016;**162-163**:129-133. DOI: 10.1016/j.jenvrad.2016.05.006

[39] Olasupo FO, Olumuyiwa Ilori C, Forster BP, Bado S. Mutagenic effects of gamma radiation on eight accessions of Cowpea (*Vigna unguiculata* [L.] Walp.). American Journal of Plant Sciences. 2016;**07**(02):339-351. DOI: 10.4236/ajps.2016.72034

[40] Antúnez-Ocampo OM, Cruz-Izquierdo S, Mendoza-Onofre LE,

Sandoval-Villa M, Santacruz-Varela A, de La Cruz-Torres E, et al. Growth dynamics of morphological and reproductive traits of *Physalis peruviana* L. M1 plants obtained from seeds irradiated with gamma rays. Not Bot Horti Agrobot Cluj Napoca. 2020;**48**(1):200-209. DOI: 10.15835/nbha48111745

[41] Asare AT, Mensah F, Acheampong S, Asare-Bediako E, Armah J. Effects of gamma irradiation on agromorphological characteristics of okra (*Abelmoschus esculentus* L. moench.). Advanced Agriculture. 2017;**2017**:1-7. DOI: 10.1155/2017/2385106

[42] Wang P, Zhang Y, Zhao L, Mo B, Luo T. Effect of gamma rays on *Sophora davidii* and detection of DNA polymorphism through ISSR marker. BioMed Research International. 2017;**2017**:8576404. DOI: 10.1155/2017/8576404

[43] Qi W, Zhang L, Wang L, Xu H, Jin Q, Jiao Z. Pretreatment with low-dose gamma irradiation enhances tolerance to the stress of cadmium and lead in *Arabidopsis thaliana* seedlings. Ecotoxicology and Environmental Safety. 2015;**115**:243-249. DOI: 10.1016/j.ecoenv.2015.02.026

[44] Maherchandani N. Effects of gamma radiation on the dormant seed of *Avena fatua* L. Radiation Botany. 1975;**15**(4):439-443. DOI: 10.1016/0033-7560(75)90018-6

[45] Majeed A, Ahmad H, Muhammad Z. Variation in chlorophyll contents and grain yield of *Lepidium sativum* L as induced by gamma irradiation. International Journal of Biological Sciences and Engineering. 2010;**1**(2):147-151

[46] Abdel-Hady M, Okasha E, Soliman S, Talaat M. Effect of gamma radiation and gibberellic acid on germination and alkaloid production in *Atropa belladonna* L. Australian Journal of Basic and Applied Sciences. 2008;**2**(3):401-405

[47] Dubey A, Yadav J, Singh B. Studies on induced mutations by gamma irradiation in okra (*Abelmoschus esculentus* (L.) Monch.). Progressive Agriculture. 2007;**7**(1-2):46-48

[48] Moussa HR. Role of gamma irradiation in regulation of NO3 level in rocket (*Eruca vesicaria* subsp. sativa) plants. Russ. Journal of Plant Physiology. 2006;**53**:193-197

[49] Hameed M, Naz N, Ahmad M, Islam-ud-Din R. Morphological adaptations of some grasses from the salt range. Pakistan Journal of Botany. 2008;**40**(4):1571-1578

[50] Ahuja S, Kumar M, Kumar P, Gupta VK, Singhal RK, Yadav A, et al. Metabolic and biochemical changes caused by gamma irradiation in plants. Journal of Radioanalytical and Nuclear Chemistry. 2014;**300**(1):199-212. DOI: 10.1007/s10967-014-2969-5

[51] Hassanien EH, Abdeltawab FM, Elsouedy A, Sharabash MT, Mahmoud AA. Effect of gamma irradiation on growth, yield and chemical constituents for three tomato varieties and their crosses. In: El-Mashri SM, editor. Proceedings of the Second Arab Conference on the Peaceful Uses of Atomic Energy Part II: A and B. Cairo, Egypt; 1995. p. 1199

[52] Mounir AM, El-Hefny AM, Mahmoud SH, El-Tanahy AMM. Effect of low gamma irradiation doses on growth, productivity and chemical constituents of Jerusalem artichoke (*Helianthus tuberosus*) tubers. Bulletin Natural Research Center. 2022;**46**(1). DOI: 10.1186/s42269-022-00838-5

[53] Gunkel J, A. S. Ionizing radiations: biochemical, physiological and

morphological aspects of their effect on plants. Encyclopedia in Plant Physics. 1961;**16**:555-611

[54] Benedek M, Pannonhalmi K, Izsaki Z, Jeszenak G, M. M. Investigation on radioactive stimulation in tomatoes. Horticultural Abstracts. 1973;**43**:21-48

[55] Alarkon K, Bozova L, Stoeva N. Index of earliness in tomato plants produced by irradiation of seeds and transplants with gamma rates. Rastenievydni Nauki. 1987;**24**(2):40-43

[56] Voloozh D, Zham-Yansuren D. The effect of gamma irradiation of seed on the yield of outdoor tomatoes in Mongolia. Atomic Energia. 1977;**41**:149-151

[57] Kim J-H, Baek M-H, Chung BY, Wi SG, Kim J-S. Alterations in the photosynthetic pigments and antioxidant machineries of red pepper (*Capsicum annuum* L.) seedlings from gamma-irradiated seeds. Journal of Plant Biology. 2004;**47**(4):314-321. DOI: 10.1007/bf03030546

[58] Preuss SB, Britt AB. A DNA-damage-induced cell cycle checkpoint in *Arabidopsis*. Genetics. 2003;**164**(1):323-334. DOI: 10.1093/genetics/164.1.323

[59] Hong MJ, Kim DY, Jo YD, Choi H-I, Ahn J-W, Kwon S-J, et al. Biological effect of gamma rays according to exposure time on germination and plant growth in wheat. Applied Science (Basel). 2022;**12**(6):3208. DOI: 10.3390/app12063208

[60] Beckers GJM, Conrath U. Priming for stress resistance: From the lab to the field. Current Opinion in Plant Biology. 2007;**10**(4):425-431. DOI: 10.1016/j.pbi.2007.06.002

[61] van Hulten M, Pelser M, van Loon LC, Pieterse CMJ, Ton J. Costs and benefits of priming for defense in Arabidopsis. Proceedings of the National Academy of Sciences of the United States of America. 2006;**103**(14):5602-5607. DOI: 10.1073/pnas.0510213103

[62] Wang X, Ma R, Cui D, Cao Q, Shan Z, Jiao Z. Physio-biochemical and molecular mechanism underlying the enhanced heavy metal tolerance in highland barley seedlings pre-treated with low-dose gamma irradiation. Scientific Reports. 2017;**7**(1):14233. DOI: 10.1038/s41598-017-14601-8

[63] Zhang L, Zheng F, Qi W, Wang T, Ma L, Qiu Z, et al. Irradiation with low-dose gamma ray enhances tolerance to heat stress in Arabidopsis seedlings. Ecotoxicology and Environmental Safety. 2016;**128**:181-188. DOI: 10.1016/j.ecoenv.2016.02.025

[64] Mahmoud NS, Awad SH, Madani RMA, Osman FA, Elmamoun K, Hassan AB. Effect of γ radiation processing on fungal growth and quality characteristics of millet grains. Food Science & Nutrition. 2016;**4**(3):342-347. DOI: 10.1002/fsn3.295

[65] Vanhoudt N, Horemans N, Wannijn J, Nauts R, Van Hees M, Vandenhove H. Primary stress responses in *Arabidopsis thaliana* exposed to gamma radiation. Journal of Environmental Radioactivity. 2014;**129**:1-6. DOI: 10.1016/j.jenvrad.2013.11.011

[66] Hallem M. Pre-exposure to gamma rays alleviates the harmful effect of salinity on cowpea plants. Journal of Stress Physiology & Biochemistry. 2012;**8**(4):199-217

[67] Oliveira NM, de Medeiros AD, de Nogueira M, Arthur V, Mastrangelo T, da Barboza Silva C. Hormetic effects of low-dose gamma rays in soybean seeds and seedlings: A detection technique

using optical sensors. Computers and Electronics in Agriculture. 2021;**187**:106251. DOI: 10.1016/j.compag.2021.106251

[68] Singh B, Ahuja S, Singhal RK, Venu BP. Effect of gamma radiation on wheat plant growth due to impact on gas exchange characteristics and mineral nutrient uptake and utilization. Journal of Radioanalytical and Nuclear Chemistry. 2013;**298**(1):249-257. DOI: 10.1007/s10967-012-2342-5

[69] Ahuja S, Singh B, Gupta VK, Singhal RK, Venu Babu P. Very low dose gamma irradiation stimulates gaseous exchange and carboxylation efficiency, but inhibits vascular sap flow in groundnut (*Arachis hypogaea* L.). International Journal of Radiation Biology. 2014;**90**(2):179-186. DOI: 10.3109/09553002.2014.868615

[70] Cohen I, Sapir Y, Shapira M. A conserved mechanism controls translation of Rubisco large subunit in different photosynthetic organisms. Plant Physiology. 2006;**141**(3):1089-1097. DOI: 10.1104/pp.106.079046

[71] Chandorkar KR, Clark GM. Physiological and morphological responses of *Pinus strobus* L. and *Pinus sylvestris* L. seedlings subjected to low-level continuous gamma irradiation at a radioactive waste disposal area. Environmental and Experimental Botany. 1986;**26**(3):259-270. DOI: 10.1016/0098-8472(86)90038-9

[72] McCabe J, Shelp B, Ursino DJ. Photosynthesis and photo phosphorylation in radiation-stressed soybean plants and the relation of these processes to photoassimilate export. Environmental and Experimental Botany. 1979;**19**(4):253-261. DOI: 10.1016/0098-8472(79)90027-3

[73] Macovei A, Garg B, Raikwar S, Balestrazzi A, Carbonera D, Buttafava A, et al. Synergistic exposure of rice seeds to different doses of γ-ray and salinity stress resulted in increased antioxidant enzyme activities and gene-specific modulation of TC-NER pathway. BioMed Research International. 2014;**2014**:676934. DOI: 10.1155/2014/676934

[74] Ling A, Chia J, Hussein S, Harun A. Physiological responses of *Citrus sinensis* to gamma irradiation. World Applied Sciences Journal. 2008;**5**(1):12-19

[75] Stoeva N, Bineva T. Gamma irradiation treatment: Growth, photosynthesis rate and content of plastid pigments. Journal of Environmental Protection and Ecology. 2001;**2**(2):299-303

[76] Wi SG, Chung BY, Kim J-S, Kim J-H, Baek M-H, Lee J-W, et al. Effects of gamma irradiation on morphological changes and biological responses in plants. Micron. 2007;**38**(6):553-564. DOI: 10.1016/j.micron.2006.11.002

[77] Qi W, Zhang L, Feng W, Xu H, Wang L, Jiao Z. ROS and ABA signaling are involved in the growth stimulation induced by low-dose gamma irradiation in *Arabidopsis* seedling. Applied Biochemistry and Biotechnology. 2015;**175**(3):1490-1506. DOI: 10.1007/s12010-014-1372-6

[78] Gomes T, Xie L, Brede D, Lind O-C, Solhaug KA, Salbu B, et al. Sensitivity of the green algae *Chlamydomonas reinhardtii* to gamma radiation: Photosynthetic performance and ROS formation. Aquatic Toxicology. 2017;**183**:1-10. DOI: 10.1016/j.aquatox.2016.12.001

[79] Ashraf M, Cheema A, Rashid M, Qamar Z. Effect of gamma rays on M1 generation in basmati rice. Pakistan Journal. 2003;**35**(5):791-795

[80] Kreslavski VD, Los DA, Allakhverdiev SI, Kuznetsov VV.

Signaling role of reactive oxygen species in plants under stress. Russian Journal of Plant Physiology. 2012;**59**(2):141-154. DOI: 10.1134/s1021443712020057

[81] El-Beltagi HS, Ahmed OK, El-Desouky W. Effect of low doses γ-irradiation on oxidative stress and secondary metabolites production of rosemary (*Rosmarinus officinalis* L.) callus culture. Radiative Physics and Chemical Oxford England 1993. 2011;**80**(9):968-976. DOI: 10.1016/j.radphyschem.2011.05.002

[82] Gicquel M, Taconnat L, Renou J-P, Esnault M-A, Cabello-Hurtado F. Kinetic transcriptomic approach revealed metabolic pathways and genotoxic-related changes implied in the Arabidopsis response to ionising radiations. Plant Science. 2012;**195**:106-119. DOI: 10.1016/j.plantsci.2012.06.015

[83] Vardhan PV, Shukla LI. Gamma irradiation of medicinally important plants and the enhancement of secondary metabolite production. International Journal of Radiational Biology. 2017;**93**(9):967-979. DOI: 10.1080/09553002.2017.1344788

[84] Aly A, Eliwa N, AbdEl-Megid M. Stimulating effect of gamma radiation on some active compounds in eggplant fruits. Egypt Journal of Radiative Science Applications. 2019;**2019**:61-63. DOI: 10.21608/ejrsa.2019.10024.1066

[85] Kim S-H, Song M, Lee KJ, Hwang S-G, Jang CS, Kim J-B, et al. Genome-wide transcriptome profiling of ROS scavenging and signal transduction pathways in rice (*Oryza sativa* L.) in response to different types of ionizing radiation. Molecular Biology Reports. 2012;**39**(12):11231-11248. DOI: 10.1007/s11033-012-2034-9

[86] Kryvokhyzha MV, Krutovsky KV, Rashydov NM. Differential expression of flowering genes in *Arabidopsis thaliana* under chronic and acute ionizing radiation. International Journal of Radiational Biology. 2019;**95**(5):626-634. DOI: 10.1080/09553002.2019.1562251

[87] Kozeko L, Talalaiev O, Neimash V, Povarchuk V. A protective role of HSP90 chaperone in gamma-irradiated *Arabidopsis thaliana* seeds. Life Science Space Research (Amst). 2015;**6**:51-58. DOI: 10.1016/j.lssr.2015.07.002

[88] Kim J-H, Ryu TH, Lee SS, Lee S, Chung BY. Ionizing radiation manifesting DNA damage response in plants: An overview of DNA damage signaling and repair mechanisms in plants. Plant Science. 2019;**278**:44-53. DOI: 10.1016/j.plantsci.2018.10.013

[89] Kim SW, Jung IJ, Kim SH, Choi H-I, Kang S-Y, Kim J-B. Physiological and molecular analysis of OsTPS30 by gamma irradiation. Journal of Plant Biotechnology. 2019;**46**(2):88-96. DOI: 10.5010/jpb.2019.46.2.088

[90] Chang S, Lee U, Hong MJ, Jo YD, Kim J-B. High-throughput phenotyping (HTP) data reveal dosage effect at growth stages in *Arabidopsis thaliana* irradiated by gamma rays. Plants. 2020;**9**(5):557. DOI: 10.3390/plants9050557

[91] Rezk AA, Al-Khayri JM, Al-Bahrany AM, El-Beltagi HS, Mohamed HI. X-ray irradiation changes germination and biochemical analysis of two genotypes of okra (*Hibiscus esculentus* L.). Journal of Radiation Research and Applied Sciences. 2019;**12**(1):393-402. DOI: 10.1080/16878507.2019.1680188

[92] Dada KE, Animasaun DA, Mustapha OT, Bado S, Foster BP. Radiosensitivity and biological effects of gamma and X-rays on germination

and seedling vigour of three *Coffea arabica* varieties. Journal of Plant Growth Regulation. 2022;**2022**:1582-1591. DOI: 10.1007/s00344-022-10643-z

[93] Arena C, De Micco V, De Maio A. Growth alteration and leaf biochemical responses in *Phaseolus vulgaris* exposed to different doses of ionising radiation. Plant Biology (Stuttgart, Germany). 2014;**16**(Suppl. 1):194-202. DOI: 10.1111/plb.12076

[94] Deoli NT, Hasenstein KH. Irradiation effects of MeV protons on dry and hydrated *Brassica rapa* seeds. Life Science Space Research (Amst). 2018;**19**:24-30. DOI: 10.1016/j.lssr.2018.08.004

[95] Im J, Kim WJ, Kim SH, Ha B-K. Effects of proton beam irradiation on seed germination and growth of soybean (*Glycine max* L). Journal of Korean Physical Society. 2017;**71**(11):752-757. DOI: 10.3938/jkps.71.752

[96] Kim S-K, Park S-Y, Kim K-R, Shin J-H, Kim S-Y, Kim H-Y, et al. Effect of proton beam irradiation on germination, seedling growth, and pasting properties of starch in rice. Journal of Crop Science and Biotechnology. 2012;**15**(4):305-310. DOI: 10.1007/s12892-012-0063-5

[97] Kumar V, Vishwakarma G, Chauhan A, Shitre A, Da BK, Nair J, et al. Use of proton beam as a novel tool for mutations in rice. BARC Newsletter. 2018;**366**:5-9

[98] Lung H-M, Cheng Y-C, Chang Y-H, Huang H-W, Yang BB, Wang C-Y. Microbial decontamination of food by electron beam irradiation. Trends in Food Science and Technology. 2015;**44**(1):66-78. DOI: 10.1016/j.tifs.2015.03.005

[99] Waskow A, Butscher D, Oberbossel G, Klöti D, Rudolf von Rohr P, Büttner-Mainik A, et al. Low-energy electron beam has severe impact on seedling development compared to cold atmospheric pressure plasma. Scientific Reports. 2021;**11**(1):16373. DOI: 10.1038/s41598-021-95767-0

[100] Palomino G, Nepamuceno F, Villalobos-Pietrini R. A general description of barley coleoptile growth behavior under low LET radiations. Environmental and Experimental Botany. 1979;**19**(2):105-115. DOI: 10.1016/0098-8472(79)90015-7

[101] Doroshkevich SY, Artemov KP, Tereshchenko NN, Zyubanova TI, Vorobyov MS, Akimova EE, et al. Presowing treatment of spring wheat seeds by a pulsed electron beam in the atmosphere. High Energy Chemistry. 2021;**55**(4):329-335. DOI: 10.1134/s0018143921040068

[102] Bourke P, Ziuzina D, Boehm D, Cullen PJ, Keener K. The potential of cold plasma for safe and sustainable food production. Trends in Biotechnology. 2018;**36**(6):615-626. DOI: 10.1016/j.tibtech.2017.11.001

[103] Mildaziene V, Ivankov A, Sera B, Baniulis D. Biochemical and physiological plant processes affected by seed treatment with non-thermal plasma. Plants. 2022;**11**(7):856. DOI: 10.3390/plants11070856

[104] Meiqiang Y, Mingjing H, Buzhou M, Tengcai M. Stimulating effects of seed treatment by magnetized plasma on tomato growth and yield. Plasma Science and Technology. 2005;**7**(6):3143-3147. DOI: 10.1088/1009-0630/7/6/017

[105] Dubinov AE, Lazarenko ER, Selemir VD. Effect of glow discharge air plasma on grain crops seed. IEEE Transactions on Plasma Science IEEE Nuclear Plasma Science

Society. 2000;**28**(1):180-183. DOI: 10.1109/27.842898

[106] Será B, Stranák V, Serý M, Tichý M, Spatenka P. Germination of *Chenopodium album* in response to microwave plasma treatment. Plasma Science and Technology. 2008;**10**(4):506-511. DOI: 10.1088/1009-0630/10/4/22

[107] Šerá B, Šerý M, Štraňák V, Špatenka P, Tichý M. Does cold plasma affect breaking dormancy and seed germination? A study on seeds of lamb's quarters (*Chenopodium album* agg.). Plasma. Science and Technology. 2009;**11**(6):750-754. DOI: 10.1088/1009-0630/11/6/22

[108] Gholami A, Safa NN, Khoram M, Hadian J, Ghomi H. Effect of low-pressure radio frequency plasma on ajwain seed germination. Plasma Medicine. 2016;**6**(3-4):389-396. DOI: 10.1615/plasmamed.2017019157

[109] Măgureanu M, Sîrbu R, Dobrin D, Gîdea M. Stimulation of the germination and early growth of tomato seeds by non-thermal plasma. Plasma Chemistry and Plasma Processing. 2018;**38**(5):989-1001. DOI: 10.1007/s11090-018-9916-0

[110] Štěpánová V, Slavíček P, Kelar J, Prášil J, Smékal M, Stupavská M, et al. Atmospheric pressure plasma treatment of agricultural seeds of cucumber (*Cucumis sativus* L.) and pepper (*Capsicum annuum* L.) with effect on reduction of diseases and germination improvement. Plasma Processes and Polymers. 2018;**15**(2):1700076. DOI: 10.1002/ppap.201700076

[111] Figueira FR, Doria ACOC, Khouri S, Maciel HS, Pessoa RS, Ramos MAR. Effect of storage temperature on pH and conductivity of reverse osmosis water treated with atmospheric plasma. Plasma Medicine. 2018;**8**(3):237-244. DOI: 10.1615/plasmamed.2018028327

[112] Thirumdas R, Kothakota A, Annapure U, Siliveru K, Blundell R, Gatt R, et al. Plasma activated water (PAW): Chemistry, physico-chemical properties, applications in food and agriculture. Trends in Food Science and Technology. 2018;**77**:21-31. DOI: 10.1016/j.tifs.2018.05.007

[113] Fan L Liu X, Ma Y, Xiang Q. Effects of plasma-activated water treatment on seed germination and growth of mung bean sprouts. Journal of Taibah University for Science. 2020;**14**(1):823-830. DOI: 10.1080/16583655.2020.1778326

[114] Kučerová K, Henselová M, Slováková Ľ, Hensel K. Effects of plasma activated water on wheat: Germination, growth parameters, photosynthetic pigments, soluble protein content, and antioxidant enzymes activity. Plasma Processes and Polymers. 2019;**16**(3):1800131. DOI: 10.1002/ppap.201800131

[115] Sperling L, McGuire S. Fatal gaps in seed security strategy. Food Security. 2012;**4**(4):569-579. DOI: 10.1007/s12571-012-0205-0

Chapter 2

Influence of Seed Development and Maturation on the Physiological and Biochemical Seed Quality

Morish Obura and Jimmy Lamo

Abstract

Seed quality is one of the widely discussed topics in seed system and seed biology; thus, many countries with functional and vibrant seed system have invested heavily in seed quality assurance and quality control. Good quality seed is crucial for any cropping system, for without it, there is poor field establishment and wastage of other production inputs. Good quality seed responds well to added inputs, ensures uniform crop establishment, and has higher yield advantage to poor quality seed under the same management practice. It is, however, important to note that seed quality is influenced greatly by seed development and maturation. Storage reserves are deposited in seed storage tissues during seed development and maturation, and these reserves are important in the early stages of germination and maintenance of seedling life when it has not yet developed good photosynthetic capacity. The development stage at which the seed is harvested has enormous influence on its performance either in the field or storage, in terms of germination behavior and vigor characteristics, and maintenance of viability. This chapter presents some of the current understandings and findings on seed development and maturation, with emphasis on the physiological and biochemical quality.

Keywords: seed development, seed maturation, physiological seed quality, biochemical seed quality, physiological maturity

1. Introduction

Seed quality, the standard of excellence in the seed characteristics is what determines its performance when sown or stored [1]. Seed quality has been recognized as a complex trait, and has therefore been described as the viability and vigor characteristics of the seed that allows emergence and establishment of seedlings in diverse environmental conditions [2]. The fundamental and most important input in agriculture is good quality seed [3]. The main objective of any seed system is to ensure that good quality seed is delivered to the end users who are farmers. Every country in the world has put in place regulatory measures which seed producers must adhere to, in order to reduce quality loss or adulteration of seed along the seed value chain. A case in point, Uganda through the Ministry of Agriculture, Animal Industry and Fisheries (MAAIF) enacted the national seed policy 2018 with the vision of a competitive, profitable and sustainable seed sub-sector where farmers and all seed users have access to affordable quality seed", and four strategic policy objectives of (i) strengthening research and development for the seed sector, (ii) strengthening capacity of the key players along the seed value chain to achieve an effective and efficient seed sector, (iii) strengthening the seed quality control system along the entire value chain and (iv) enhancing knowledge and information management for the seed sector [4]. Seed testing, certification, variety release and registration, phytosanitary measures and protection of plant breeders' rights are some of key activities governed by a seed law, and the management of these activities either at a country, regional or continental level affects the outcome of seed production, availability, accessibility, and hence the design of agricultural system [5].

Four seed quality attributes commonly talked about are; physical quality, pathological quality, physiological quality and genetic quality [6, 7]. Biochemical seed quality is often put together with the physiological quality although the two are different. To attain maximum seed quality, it has been argued that seed should be harvested at physiological maturity (PM). However, controversies exist in the concept of physiological maturity as explained by [8]. Some crop species maintain high seed moisture content at PM that makes harvesting difficult due to mechanical damages to the seed, hence harvesting should be delayed for some days after PM. Harvesting of seeds which are produced in fleshy fruits can be done just before PM and fruits given a period of after ripening to complete seed maturation [9]. Three common concepts of PM have been presented [10] (i) stage of maximum seed dry matter accumulation, (ii) growth stage beyond which there is no significant increase in seed dry weight and (iii) growth stage when seed attains maximum dry weight, germination and vigor. Given the existing controversies in the use of PM concept, seed development should be traced during seed development and maturation. This should involve tagging flowers at anthesis, harvesting seed or fruit at different development stages and evaluating morphological, physiological and biochemical changes in the seed, as well as correlation between these attributes in order to give a concrete judgment on when maximum seed quality is attained during seed development. This book chapter therefore looks at studies that have been done in different crop species to evaluate physiological and biochemical seed quality during seed development and maturation.

2. Seed development and maturation

Seed development is the changes in structural and physiological events in the seed right from the time of ferritization until the seed reaches maturation. A viable pollen comes in contact with the stigma, followed by its germination to form pollen tube carrying two male nuclei. The pollen tube penetrates the embryo sac containing egg nucleus and polar nuclei. The first male nucleus fuses with the egg cell to form a diploid zygote and the second one fuses with the polar nuclei to form a triploid endosperm; this is referred to as double fertilization and it marks the process of successful seed development in angiosperms [11]. Seed development is characterized in three stages; Histodifferentiation and cell expansion (stage I), Reserve deposition, cell expansion and maturation (stage II), and Maturation drying (stage III) [12, 13]. Stage I is characterized by the formation of embryonic tissues and those encompassing it, stage II is characterized by an increase in seed dry weight as a result of the accumulation of storage reserves, and physiological maturity is reached at the end of this stage. Stage III exhibits a decrease in seed dry matter as the seed reaches its mature form [12, 13]. The very prominent attribute of stage III is the acquisition of desiccation tolerance, arrest of growth, and entry into dormancy [14]. In orthodox seeds, the seed loses about 10–15% of its moisture content to develop a desiccation tolerance and remain quiescent [12]. Embryogenesis in dicot and monocot occurs with similar pattern of events but the embryos formed are structurally different [13]. Seeds develop desiccation tolerance even before attaining a physiological maturity status. Physiological maturity is the development stage of the seed when the seed attains maximum dry weight [10]. Deposition and aggregation of storage compounds such as heat shock proteins, late embryonic abundants (LEAs) proteins and antioxidants of lower molecular weights in the seed embryo during seed development and maturation are associated with the desiccation tolerance in seed [15]. Seed development is regulated by several hormones. Indole-3-acetic acid (IAA), is crucial in determination of embryo size and structure during embryogenesis [16, 17]. Abscisic acid (ABA) induces dormancy which prevents unwanted seed germination and also promotes deposition of storage reserves during seed development, and formation of late embryonic abundants (LEAs) proteins which protect the seed from desiccation [18]. There is variation in ABA peaks during seed development, for example wheat has two peaks while rice has one peak during seed development [19, 20]. Cytokinins promotes cell division and differentiation and counteracts the negative effects of ABA during seed development [21]. Gibberellic acid is another hormone which is very critical during seed development as it antagonizes the inhibitory effects of Abscisic acid, but complex relationship exist between the two hormones in different crop species during seed development. Other than hormones, several transcriptional factors form a complex network to regulate seed development and maturation events [22]. Such events are storage reserve accumulation, chlorophyll degradation, and the acquisition of primary dormancy, desiccation tolerance, and longevity [23, 24]. During seed development and maturation, physiological and biochemical changes related to seed germination, vigor, seed storage proteins, lipids, antioxidant enzymes, sugars occur in the seed, with marked variations among crop species and varieties of the same crop. When seed attains maximum dry weight, deposition of storage reserves ceases as the seed enters the late maturation phase during which seed vigor is developed and the seed builds a defensive mechanism to aid survival after dispersal from the mother plant [25–27]. Seed harvest maturity stage greatly impacts seed germination, vigor, viability, storability and longevity hence production of good quality seed is very dependent on this factor [28–30]. In an informal seed system where farmers exchange seed among themselves, no much attention is paid to the seed age at harvest because of limited knowledge on how this aspect affect seed quality. An example is the vegetable farmers who keep harvesting the fruits and extract seeds from last harvest or from those left from fruits sold in the market. This practice has resulted in marked variability in the seed quality in the informal system. Farmer training to help them understand seed development is very crucial in addressing seed quality issues that they encounter.

3. Seed development and maturation effects on physiological seed quality

Seed harvest maturity stage is one of the factors that affect physiological seed quality. Physiological maturity, often defined as a maturity stage at which the seed accumulates maximum dry matter, sometimes coincides with maximum seed germination, example in *Solanum aethiopicum* [31] and Okra [32] but in the case of species such as tomato [33], soybean [34] and pepper [35], this scenario is not applicable. Botey et al. [36] studied seed development of two African eggplant cultivars; *Oforiwa* and *Kpando* at different seed development stages of 20, 34, 48, 62, 76 and 82 days after anthesis (DAA) and reported that maximum seed dry weight was attained at 48 and 76 DAA in *Oforiwa* and *Kpando* respectively under a characteristic tropical climate, and 62 and 76 DAA respectively under a characteristic temperate oceanic climate, but maximum germination percentage only coincided with physiological maturity in cultivar *Kpando*. Another study in six eggplant cultivars; *Dwomo* and *Kpando* belonging to *Solanum gilo* group, GH 3870 and GH 3887 belonging to *Solanum melongena* group, GH 107 and GH 4918 belonging to *Solanum macrocarpon* group showed that seed germination and vigor improved when seed were harvested from 4 weeks to 8 weeks after full maturity of the fruits [31]. This study tends to suggest that maturation of seeds in fleshy fruits may not coincide with that of the fruit itself as maturation of the seeds continue even after the fruit reaches full maturity. Kwankaew et al. [37] studied the seed quality of Upland Rice cultivar *Dawk Pa-yawm* at eight different maturity stages of 8, 12, 16, 20, 24, 28, 32, and 36 days after flowering (DAF) and observed that seed physiological characteristics in terms of germination and vigor were all maximum at physiological maturity but all decreased after this stage. Accordingly, the authors reported maximum seed dry wright of 21.89 mg/seed, germination capacity (97%), soil emergenc3e (96.5%), and maximum seedling dry weight (7.51 mg/seedling), root length (13.3 cm) and shoot length (7.92 cm), and lowest electrical conductivity of 8.05 µS/cm/g at 28 DAF. Low electrical conductivity, high seedling dry weight, high root and shoot lengths are indicators of high seed vigor. In Okra cultivar *Asontem*, [32] studied five seed development stages of 10, 20, 30, 40 and 50 DAA and reported no seed germination up to 20 DAA, and maximum seed germination of 77% at 50 DAA. Santos et al. [38] studied seed development in Okra cultivar Santa Cruz 47 by tagging flowers and harvesting fruits from 5 DAA to 65DAA and extracted seeds immediately or stored for 7 days, and reported maximum seed germination, seedling emergence and germination first count at 50DAA both for seeds extracted immediately after harvest and those extracted 7 days after harvest. They observed no germination and seedling emergence up to 25 DAA for seeds extracted immediately after harvest while those extracted after 7 days of storage started germinating at 20 DAA. Their result in an indication that seeds borne in fruits continue to develop during post-harvest period when detached from the mother plant, provided they are stored in suitable conditions. Evaluation of free space percentage and aspect ratio of Okra seed during seed development using X-ray imaging analysis showed that both free space and aspect ratio decreased during seed development, stabilizing around 50DAA and were strongly linked to good germination and vigor of the seed [38]. By harvesting three soybean varieties *Nangbaar*, *Anidaso*, *Jenguma* at physiological maturity (PM), one week after PM and two weeks after PM, [39] observed 85.25%, 85.25% and 66.75% in *Nangbaa*r, *Anidaso*, *Jenguma* respectively at PM while the germination decreased to 67.33%, 60.92% and 58.83% in *Nangbaar*, *Anidaso*, *Jenguma* respectively when they were harvested two weeks after PM. The same study reported high seed vigor when seed was harvested at PM and low seed vigor at one week and two weeks after physiological maturity, as indicated by low electrical conductivity (EC) of seed at PM and high EC of seed one week and two weeks after PM. In cucumber seeds, [40] reported a maximum germination capacity of 80.84% when fruits grown under open field conditions were harvested at 40 days after flowering followed by 30 days postharvest ripening. Seed vigor characteristics such as seedling dry weight, seedling vigor index I and II were also maximum at the same fruit harvest stage under the same environment. In Bell pepper (*Capsicum annuum*), [35] studied the effect of harvesting time on seed quality of two cultivars *Fyuco INTA* and *Lungo INTA* by tagging flowers and harvesting at 4, 5, 6, 7, 8 and 9 weeks after anthesis (WAA). Their study reported very poor germination and low seed vigor in both cultivars for seeds harvested from 4 to 7 WAA both in fresh seeds and those stored for one year, while maximal seed quality in terms of seed germination and vigor was only attained in the two cultivars when seeds were harvested at 9 WAA. In sweet pepper cultivar *Amarela comprida,* [41] tagged flowers and harvested fruits from 20 DAA to 75 DAA, and reported no seed germination up to 40DAA, with maximum seed germination and seed vigor coinciding with maximum dry matter at 75DAA. In a related study in Habanero pepper (*Capsicum chinenses* Jacq), [42] harvested the green, yellow and orange fruits corresponding to 30, 38, and 42DAA and stored for 7 and 14 days and observed highest seed germination and vigor in seeds harvested from orange fruits followed by yellow while those extracted from green fruits had the least germination and vigor both under storage period of 0, 7 and 14 days. In another study, [43] evaluated the physiological seed quality of *Physalis angulata* L. seeds under different harvesting periods of 15, 22, 29, 36 and 43 DAA, and observed 100% seed germination, and maximum field emergence of 70.5%, at 29DAA, while the seed attained maximum seed dry weight of 28 mg at 36DAA indicating that maximal seed quality was attained before physiological maturity. Tetteh et al. [44] studied seed development in two tomato cultivars GH 9207 and GH 9305 and classified the fruits as initially ripe, half ripe, fully ripe and rotten depending on the development stages, and observed good seed vigor and germination in fully ripe fruits but were not statistically different from germination of those harvested from half ripe and rotten fruits. Seed should be harvested at the maturity stage when germination and vigor are maximum [45, 46]. A study evaluated seed development in pumpkin by harvesting the seed at 30, 40, 50 and 60 DAA and observed less than 20% germination for seeds harvested at 30 and 40DAA, and more than 80% for seeds harvested at 60DAA in both round type and oval type pumpkin fruits in all the three locations [30]. It was asserted that seeds of most crops attain maximum germination and vigor at PM and declines after [47], but this concept of PM has been debated by many authors including [8, 10]. Crop species differ in their seed and fruit maturation characteristics, hence the maturity stage at which seed attain maximum germination and vigor varies. This phenomenon has been demonstrated, for example maximum germination and vigor in tomato seeds were attained 15 days after PM [48] while [49] reported 20 days after PM in the same crop, and [50] reported 10 days after PM in pepper. Changes in fruit color, fruit weight, fruit diameter and length, seed dry weight and seed moisture content during seed development can be used as indicators for seed maturation and harvesting to attain good seed quality [36, 42, 44]. Seed vigor during seed development can be measured indirectly using electrical conductivity, with lower EC values indicating high seed vigor and vice versa. Seed vigor is low at initial stages of seed development, as indicated by very high electrical conductivity but increases as the seed matures due to strengthening of seed membrane integrity that reduces the leakages of electrolytes from the seed. This has been verified during seed development in *Capsicum baccatum* [51], onion [52], *C. annuum* [41] and faba bean [53]. No seedling emergence was observed in *C. baccatum* seeds until 30 DAA, but emergence increased from 39.5% at 40 DAA to 73.5% at 50 DAA and remained statistically unchanged until last harvest at 80 DAA [51]. Several studies on seed development and maturation have revealed that seeds of some species attain germination potential very early just few days after anthesis but some requires sometime in order to attain the ability to germinate. This variation exists even within the same species as observed in two species of *S. aethiopicum, Oforiwa* and *Kpando* [36]. However, as discussed previously, it is also possible to harvest seeds one or two weeks before physiological maturity and give them a period of after ripening to attain full maturity and hence good germination and vigor, particularly for those borne in fleshy fruits [54]. Seed development and maturation, and hence attaining of maximum seed quality is also influenced by environment and genotype, thus recommendations for seed production should capture genetic variations as well as conditions in the production environment. In addition to that, morphological changes in the fruit and seed should be related to the seed quality in terms of seed germination and vigor so that seed producers can know the best time for harvesting seed with the highest quality.

4. Changes in biochemical seed quality during seed development and maturation

A number of biochemical changes related to structural proteins, carbohydrates, lipids, antioxidant enzymes, phytic acid and tannins occur during occur in the seed during seed development and maturation. Changes in some of these compounds are related to acquisition of desiccation tolerance in the seeds and maintenance of cell membrane integrity which improves seed vigor and longevity [55, 56]. Maximum accumulation of seed proteins, and activity of alpha-amylase and dehydrogenase enzyme at 28DAA was observed in Prosso millet (*Panicum miliaceum* L) [57]. Antioxidant enzyme activity of sweet pepper (*C. annuum* L) cultivar Florinis NS 700 increased during seed development from 410 μg/g FW at 10 DAA reaching a maximum of 1550 μg/g FW at 80 DAA, with total seed phenolics following the same trend [58]. Similar results were obtained by earlier study that reported higher antioxidant activity in sweet pepper seeds extracted from mature fruits in comparison to those extracted from immature fruits [59]. However, this seems to be affected by genotype as [60, 61] obtained contrasting results in antioxidant enzyme activity trends in the same crop. High antioxidant enzyme activity during early seed development has been reported in African eggplant [62] and *Hevea brasiliensis* L [63]. Both authors reported a decline after an early increase, and increase at the end of the seed maturation phase. Antioxidant enzymes such as catalase (CAT), superoxide dismutase (SOD), peroxidase (POX) and ascorbate peroxidase (APX) protect the seed cells from oxidative damage caused by reactive oxygen species [64]. Accumulation of structural proteins such as late embryonic abundants (LEAs) has been reported to be high at 60 DAA in sweet pepper [41]. A study in *S. melongena* revealed that the levels of CAT, SOD and POX antioxidant enzymes were high in seeds harvested at 40 DAA with a progressive decrease up to 60DAA (maturation stage) and an increase after this period [65]. A similar observation in the activity of CAT and SOD was made in sweet pepper, with a decrease in their activity until maturation and an increase thereafter [66]. High respiration in younger fruits results in the production of oxidative reactive oxygen species which triggers the antioxidant enzyme system to protect the cells, hence high levels and activity of the enzymes [67, 68]. Changes in tannins content during seed development has been reported as an indicator for attaining good quality seed, as [62] observed highest tannins content in seeds of *solanum aethiopicum* cultivar *Oforiwa* harvested at physiological maturity (62DAA) and was strongly correlated with germination percentage, germination index and mean daily germination. Tannins is an antioxidant and thus improves cell wall integrity and protects it from degradation. Condensed tannins, a group of flavonoids were observed to increase throughout seed development in common bean and decreased after seed maturation [69]. A study in *Magnolia zenii* Cheng revealed a significant increase in soluble sugar, protein and lipid content of seed between 30 to 165 DAF [70]. In *capsicum baccatum*, [51] observed that soluble protein content increased by 66% from 10 to 40 DAA and remained unchanged until last harvest at 80DAA while neutral lipids content increased 14-fold between 10 and 30 DAA, reduced from 30 to 60 DAA and later increased until 80 DAA. The same authors reported that starch content was very high at the beginning of seed development, but decreased until 30DAA, and later increased until 60DAA, and remained unchanged thereafter until last harvest at 80 DAA. Starch serves as a temporary carbon storage during early seed development and constitutes to seed reserve biosynthesis [71, 72]. Non reducing sugars were observed to increase throughout seed development until 60 DAA with a slight decline until 80 DAA in *capsicum baccatum* seeds while total soluble sugars and total free amine acids showed opposite trend, but were very high at the beginning of seed development [51]. As seed undergoes desiccation during late maturation, a considerable amount of water is lost which has a potential to cause damage of cell membrane and thus affecting its stability. However, nonreducing sugars binds to the hydrophilic heads of the membrane lipids to replace the lost water and hence stabilize the cell membrane of the seed [73]. Biochemical changes during seed development have been used as markers for determination of physiological maturity and seed harvest maturity stage [51, 52, 74]. Antioxidant enzyme activity, electrical conductivity, accumulation of tannins, total sugars, total proteins and lipids have been widely used to characterize seed development and maturation in a number of crops. Using metabolites accumulation, storage reserve and seed dry weight, and moisture loss during seed development of *capsicum baccatum*, [51] were able to recognize the three seed development stages of histodifferentiation, reserve deposition and maturation drying. A study evaluated three *S. melongena* L. varieties ie Serbian variety, Italian variety and Chinese variety, through seed ripening phases of commercial ripeness, semi ripeness and full ripeness [75]. The authors reported increase in seed proteins during seed development with the highest protein accumulation at the full ripeness stage corresponding to 75, 90 and 110 DAA in Serbian, Italian and Chinese variety respectively. The same seed development stage also had highest seed germination for the three varieties. Phytic acid, the seed storage reserve of phosphorus [76] is another important compound for early seedling growth. Phytic acid binds with metallic cations to form phytate which is hydrolyzed by phytase enzymes during germination to release inorganic phosphorus and other mineral elements [77–79]. About 30–80% of seed phosphorus reserve is stored in the form of phytate [80]. High seed phosphorus content is reported to have a strong positive and significant correlation with the seed vigor [81]. A slight decrease of inorganic phosphate (Pi) throughout seed development up to 30 DAA was observed in rice both in the endosperm and aleurone layers while phytic acid content showed an increasing trend particularly in the aleurone layers peaking at 30 DAA [76], suggesting assimilation of Pi into phytic acid during seed development and maturation. In Bambara groundnut, [82] observed a gradual increase in seed phosphorus concentration from 14 DAA to 42 DAA and a sharp increase up to 62 DAA in all the four landraces evaluated. The authors also reported a strong correlation between seed phosphorus content and phytic acid. Similarly, a strong correlation between seed phosphorus concentration and phytic acid was reported in chickpea [83] and soybean [84]. Other studies have also shown that phytic acid accumulate in the seed during maturation phase in cowpea [85], chick pea [86] and mungbean [87]. Up to date, fewer studies have related biochemical markers to physiological seed quality. Accumulation of seed metabolites and storage reserves that constitute seed quality have not been well correlated with physiological state of the seed. Relating biochemical markers with morphological and physiological markers is important to give a strong justification on the seed harvest maturity period when maximal seed quality can be attained. The seed storage reserves (phytochemicals) play very important roles during seed germination when the seedling is not receiving any external input and has low photosynthetic capacity. These reserves are hydrolyzed upon reactivation of metabolic enzymes and are channeled to the growing regions of the seed during germination, thus constituting to seed vigor and establishment of strong and healthy seedlings.

5. Conclusion

Understanding of seed development and maturation is very important in producing good quality seeds in a seed system. Harvesting of seed before physiological maturity results in poor quality seeds due to immature embryo while very late harvesting results in seed aging and deterioration which lowers seed quality. However, controversies exist among several authors on the concept of physiological maturity. Some authors argued that seed harvesting at PM is not practical and economical in some species which maintain high seed moisture content at PM, thus they suggest that seed harvesting should be delayed for some time after PM to reduce seed damage especially for mechanical harvesting. For seeds borne in fleshy fruits, seed maturation can continue during postharvest ripening which improves seed quality. Several markers of physiological maturity including maximum dry matter accumulation, fruit color, leaf color and seed moisture content have been identified. To effectively study seed development, flowers should be tagged at anthesis and fruit or seed should be harvested at different times to trace seed germination and vigor characteristics, and accumulation of seed storage compounds and phytochemicals during seed development stages. However, several other factors other than seed development and maturation affect seed quality. This includes after ripening, production environment and nutrition of the mother plant, post-harvest management related to storage, and drying. During seed production, good management is required to reduce stress on the mother plant especially during seed filling, as most metabolites and phytochemicals that contribute to seed quality are accumulated in the seed during this stage. After seed filling, no more accumulation of dry matter occurs in the seed, thus irrigation and fertilizer application are not necessary. Seed development and maturation is a complex process involving many phytohormones, genes as well as transcriptional factors that play various roles at every development stage right after pollination, to development of embryo and other essential seed structures, accumulation of storage reserves and attaining physiological maturity, maturation and development of desiccation tolerance.

Author details

Morish Obura* and Jimmy Lamo
National Agricultural Research Organisation (NARO), National Crops Resources Research Institute (NaCRRI), Kampala, Uganda

*Address all correspondence to: oburamorish100@gmail.com

References

[1] Hampton JG. What is seed quality? Seed Science and Technology. 2002;**30**(1):1-10

[2] Khan N, Kazmi RH, Willems LA, Heusden VAW, Ligterink W, Hilhorst HW. Exploring the natural variation for seedling traits and their link with seed dimensions in tomato. PLoS One. 2012;**7**(8):1-14

[3] Sundareswaran S, Choudhury PR, Vanitha C, Yadava DK. Seed quality: Variety development to planting—An overview. In: Malavika D, Devendra KY, editors. Seed Science and Technology: Biology, Production, Quality. New York, USA: Springer; 2023. pp. 1-15

[4] Ministry of Agriculture, Animal Industry and Fisheries (MAAIF). Entebbe, Uganda: National Seed Policy; 2018. pp. 1-22

[5] Munyi P. Current developments in seed laws harmonisation in Africa. Report to the European Commission. DeSIRA-LIFT. 2022;**2022**:1-40

[6] Copeland LO, McDonald MF. Principles of Seed Science and Technology. New York, USA: Springer Science and Business Media; 2012

[7] Huda MN. Why quality seed? Reality & vision Bangladesh context. Evergreen Printing & Packaging. 2001;**2001**:9-156

[8] Ellis RH. Temporal patterns of seed quality development, decline, and timing of maximum quality during seed development and maturation. Seed Science Research. 2019;**29**(2):135-142

[9] Dias DCFS, Ribeiro FP, Dias LAS, Silva DJH, Vidigal DS. Tomato seed quality in relation to fruit maturation and post-harvest storage. Seed Science and Technology. 2006;**34**:691-699

[10] Bareke T. Biology of seed development and germination physiology. Advances In Plants and Agriculture Research. 2018;**8**(4):336-346

[11] Raghavan V. Double Fertilization, Embryo and Endosperm Development in Flowering Plants. Berlin: Springer; 2005. p. 234

[12] Bewley JD, Nonogaki H. Seed Maturation and Germination. New York, USA: Springer; 2017. pp. 1-9

[13] Bewley JD, Bradford KJ, Hilhorst HWM, Nonogaki H. Seeds: Physiology, Development, Germination and Dormancy. Third ed. New York: Springer; 2013

[14] Kozaki A, Aoyanagi T. Molecular aspects of seed development controlled by gibberellins and abscisic acids. International Journal of Molecular Science. 2022;**1876**:23

[15] Sripathy KV, Groot SPC. Seed development and maturation. In: Malavika D, Devendra K, editors. Seed Science and Technology: Biology, Production, Quality. New York, USA: Springer; 2023. pp. 17-38

[16] Chiwocha S, Von Aderkas P. Endogenous levels of free and conjugated forms of auxin, cytokinins and abscisic acid during seed development in Douglas fir. Plant Growth Regulation. 2002;**36**(3):191-200

[17] Locascio A, Roig-Villanova I, Bernardi J, Varotto S. Current perspectives on the hormonal control of seed development in Arabidopsis and

maize: A focus on auxin. Frontiers Plant Science. 2014;**5**(412):1-22

[18] Delahaie J, Hundertmark M, Bove J. LEA polypeptide profiling of recalcitrant and orthodox legume seeds reveals ABI3-regulated LEA protein abundance linked to desiccation tolerance. Journal of Experimental Botany. 2013;**64**(14):4559-4573

[19] Tuan PA, Kumar R, Rehal PK, Toora PK, Ayele BT. Molecular mechanisms underlying Abscisic acid/gibberellin balance in the control of seed dormancy and germination in cereals. Frontiers in Plant Science. 2018;**9**(668):1-14

[20] Liu Y, Fang J, Xu F, Chu J, Yan C, Schläppi MR, et al. Expression patterns of ABA and GA metabolism genes and hormone levels during rice seed development and imbibition: A comparison of dormant and non-dormant rice cultivars. Journal of Genetics and Genomics. 2014;**41**:327-338

[21] Matilla AJ. Auxin: Hormonal signal required for seed development and dormancy. Plants. 2020;**9**(6):705. DOI: 10.3390/plants9060705

[22] Alizadeh M, Hoy R, Lu B, Song L. Team effort: Combinatorial control of seed maturation by transcription factors. Current Opinion in Plant Biology. 2021;**63**:102091. DOI: DOI.10.1016/j.pbi.2021.102091

[23] Nonogaki H. ABA responses during seed development and germination. Advances in bbotanical research. 2019;**2019**(92):171-217

[24] Baud S, Dubreucq B, Miquel M, Rochat C, Lepiniec L. Storage reserve accumulation in Arabidopsis: Metabolic and developmental control of seed filling. Arabidopsis Book. 2008;**6**:e0113

[25] Groot SPC. Seed maturation and its practical implications. Seed Science and Technology. 2022;**50**(1):141-151

[26] Sano N, Rajjou L, North HM, Debeaujon I, Marion-Poll A, Seo M. Staying alive: Molecular aspects of seed longevity. Plant and Cell Physiology. 2015;**57**:660-674

[27] Leprince O, Pellizzaro A, Berriri S, Buitink J. Late seed maturation: Drying without dying. Journal of Experimental Botany. 2017;**68**:827-841

[28] Kole S, Gupta K. The timing of physiological maturity of seeds of sunflower. Evaluation through multiple tests. Seed Science and Technology. 2002;**10**:457-467

[29] Laercio JS, Denise FS, Carla CM, Luiz- Antonio SD. Relationship between fruit maturation stage and physiological quality of physic nut *(Jatropha curcas L.)* seeds. Agricultural Sciences. 2012;**36**(1):1413-7054

[30] Tarus WJ, Ochuodho JO, Rop NK. Influence of harvesting stage on seed quality aspects of pumpkin (*Cucurbita pepo* L.). Journal of Experimental Agriculture International. 2017;**18**(2):1-9

[31] Tetteh R, Aboagye LM, Boateng SK, Darko R. Seed quality of six eggplant cultivars as influenced by harvesting time. Journal of Applied Horticulture. 2021;**23**(1):24-27

[32] Bortey HM, Dzomeku BM. Fruit and seed quality of okra [*Abelmoschus esculentus* (L.) Moench] as influenced by harvesting stage and drying method. Indian Journal of Agricultural Research. 2016;**50**(4):330-334

[33] Borges SRS, Silva PP, Araújo FS, Souza FFJ, Nascimento WM. Tomato seed image analysis during the

maturation. Journal of Seed Science. 2019;**41**(1):022-031

[34] Zanakis GN, Ellis RH, Summerfield RJ. Seed quality in relation to seed development and maturation in three genotypes of soyabean (*Glycine max*). Experimental Agriculture. 1994;**30**:157-170

[35] Ruiz MB, Parera CA. Effect of harvesting time on seed quality of two bell pepper cultivars (*Capsicum annuum*). Revista de la Facultad de Ciencias Agrarias. 2017;**2017**:67-77

[36] Botey HM, Ochuodho JO, Ngode L. Fruit and seed physiological quality changes during seed development and maturation in African eggplant (*Solanum aethiopicum* L.). African Journal of Agricultural Research. 2021b;**17**(8):1131-1143

[37] Kwankaew T, Santipracha Q, Santipracha W. Seed development and maturation on seed quality of upland Rice cv. Dawk Pa-yawm. Walailak Journal of Science & Technology. 2017;**14**(7):607-614

[38] Santos RF, Gomes-Junior FG, Marcos-Filho J. Morphological and physiological changes during maturation of okra seeds evaluated. Scientia Agricola. 2020;**77**(3):1-9

[39] Osei TI, Banful BK, Amoah S, Apuri S, Amomba ES. Effect of harvesting stages on seed quality characteristics of three soybean (Glycine Max (L) Merrill) varieties. Journal of Scientific and Engineering Research. 2016;**3**(4):326-333

[40] Gupta N, Kumar S, Jain SK, Tomar BS, Singh J, Sharma V. Effects of stage of harvest and post-harvest ripening of fruits on seed yield and quality in cucumber grown under open field and protected environments. International Journal of Current Microbiology and Applied Sciences. 2021;**10**(1):2119-2134

[41] Vidigal DS, Dias DCFS, Dias LAS, Finger FL. Changes in seed quality during fruit maturation of sweet pepper. Scientia Agricola. 2011;**68**(5):535-539

[42] Medeiros AD, León Z, Laércio MJ, Silva J, Oliveira AMS, Dias DCFS. Relationship between internal morphology and physiological quality of pepper seeds during fruit maturation and storage. Agronomy Journal. 2020;**112**(1):1-11

[43] Ramos AR, Bassegio D, Nakagawa J, Maurício DZMD. Harvest times and seed germination of three safflower genotypes. Ciência Rural. 2021;**51**(5):1-7

[44] Tetteh R, Aboagye LM, Darko R, Osafo EA. Effect of maturity stages on seed quality of two tomato accessions. African Crop Science Journal. 2018;**26**(2):237-244

[45] Welbaum GE. Cucurbit seed development and production. HortTechnology. 1999;**9**(3):341-348

[46] Patrick JW, Offler CE. Compartmentation of transport and transfer events in developing seeds. Journal of Experimental Botany. 2001;**52**:551-564

[47] Harrington JF. Seed storage longevity. In: Kozlowski TT, editor. Seed Biology. Vol. 3. New York: Academic Press; 1972. pp. 145-245

[48] Kwon OS, Bradford KJ. Tomato seed development and quality as influenced by preharvest treatment with ethephon. HortScience. 1987;**22**:588-591

[49] Demir I, Ellis RH. Changes in seed quality during seed development and

maturation in tomato. Seed Science Research. 1992;**2**(2):81-87

[50] Oliveira AP, Gonçalves CP, Bruno RLA, Alves EU. Physiological maturation of pepper seeds in relation to fruit age after anthesis. Revista Brasileira de Sementes. 1999;**21**:88-94

[51] Silva MIL, Voigt EL, Grangeiro LC, Cunha EE, Macêdo CEC, Torres SB. Determination of harvest maturity in Capsicum baccatum L. seeds using physiological and biochemical markers. Australian Journal of Crop Science. 2015;**9**(11):1010-1015

[52] Ramya M, Yogeesha HS, Bhanuprakash K, Gowda RV. Physiological and biochemical changes during seed development and maturation in onion (*Allium cepa* L.). Vegetable Science. 2012;**39**:157-160

[53] Ghassemi-Golezani K, Hosseinzadeh-Mahootchy A. Changes in seed vigour of faba bean (*Vicia faba* L.) cultivars during development and maturity. Seed Science and Technology. 2009;**37**:713-720

[54] Passam HC, Theodoropoulou S, Karanissa T, Karapanos IC. Influence of harvest time and after-ripening on seed quality of eggplant. Scientia Horticulturae. 2010;**125**(3):518-520

[55] Weber H, Borisjuk L, Wobus U. Molecular physiology of legume seed development. Annual Review of Plant Biology. 2005;**56**:253-279

[56] Carvalho SIC, Ribeiro CSC, Henz GP. "BRS Mari": New cultivar of chili pepper for processing. Horticultura Brasileira. 2009;**27**:571-573

[57] Sridevi R, Manonmani V. Predicting the optimal stage of maximum seed quality during seed development and maturation in proso millet (*Panicum miliaceum* L). International Journal of Farm Sciences. 2019;**9**(4):89-93

[58] Papathanasiou T, Gougoulias N, Karayannis VG, Kamvoukou CA. Investigation of the Total phenolic content and antioxidant capacity of three sweet pepper cultivars (*Capsicum annuum* L.) at different development and maturation stages. Periodica Polytechnica Chemical Engineering. 2021;**65**(2):219-228

[59] Howard LR, Talcott ST, Brenes CH, Villalon B. Changes in phytochemical and antioxidant activity of selected pepper cultivars (*Capsicum* species) As influenced by maturity. Journal of Agricultural and Food Chemistry. 2000;**48**(5):1713-1720

[60] Deepa N, Kaur C, George B, Singh B, Kapoor HC. Antioxidant constituents in some sweet pepper (*Capsicum annuum* L.) genotypes during maturity. LWT – Food Science and Technology. 2007;**40**(1):121-129

[61] Ghasemnezhad M, Sherafati M, Payvast GA. Variation in phenolic compounds, ascorbic acid and antioxidant activity of five coloured bell pepper (*Capsicum annum*) fruits at two different harvest times. Journal of Functional Foods. 2011;**3**(1):44-49

[62] Botey HM. Environment, Physiological and Biochemical Effects on Seed Germination Characteristics of African Eggplant (*Solanum aethiopicum* L.). Eldoret, Kenya: University of Eldoret; 2022

[63] De Souza GA, Dias DCFS, Pimenta TM. Morpho-anatomical, physiological and biochemical changes in rubber tree seeds. Anais da Academia Brasileira de Ciências. 2018;**90**(2):1625-1641

[64] Saisanthosh K, Sumalatha GM, Shuba AC, Komala NT, Patil NK. Role of enzymatic antioxidants defense system in seeds. International Journal of Current Microbiology and Applied Sciences. 2018;**7**:584-594

[65] Aguilar AS, Cardoso AII, Vasque H, Bardiviesso EM, Felito RA, Bezerra BKL, et al. Physiological quality and antioxidant enzymes activity in eggplant seeds with different ages and resting periods after harvest. Horticultura Brasileira. 2023;**41**:e2478. DOI: 10.1590/s0102-0536-2023-e2478

[66] Colombari LF, Silva GF, Chamma L, Chaves PPN, Martins BNM, Jorge LG, et al. Maturation and resting of sweet pepper fruits on physiology quality and biochemical response of seeds. Brazilian Archives of Biology and Technology. 2021;**64**:e21200733

[67] El-Maarouf-Bouteau H, Bailly C. Oxidative signaling in seed germination and dormancy. Plant Signal Behavior. 2008;**3**:175-182

[68] Araújo RF, Abud HF, Silva LJ, Araújo EF, Pinto CMF, Silva FWS. Physiological changes and antioxidant enzymes activity in Biquinho and Malagueta pepper seeds during the maturation process. Revista Ceres. 2018;**65**(6):534-545

[69] Bett KE, Elsadr H, Marles MAS, Caldas GV, Blair MW. Condensed tannin accumulation during seed coat development in five common bean genotypes. Crop science. 2015;**55**:2826-2832

[70] Wang S, Shen Y, Bao H. Morphological, physiological and biochemical changes in Magnolia zenii Cheng seed during development. Physiologia Plantarum. 2021;**172**:2129-2141

[71] Saldivar X, Wang YJ, Chen P, Hou A. Changes in chemical composition during soybean seed development. Food Chemistry. 2011;**124**:1369-1375

[72] Djemel N, Guedon D, Lechevalier A, Salon C, Miquel M, Prosperi JM, et al. Development and composition of the seeds of nine genotypes of the *Medicago truncatula* species complex. Plant Physiology and Biochemistry. 2005;**43**:557-566

[73] Bewley JD, Bradford KJ, Hilhorst HWM, Nonogaki H. Seeds: Physiology of Development, Germination and Dormancy. 3rd ed. New York: Springer; 2012

[74] Oliveira GE, Pinho RG, Andrade T, Pinho EVR, Santos CD, Veiga AD. Physiological quality and amylase enzyme expression in maize seeds. Cienc Agrotec. 2013;**37**:40-48

[75] Popović V, Lekić S, Kiprovski B, Takač A. The effect of ripeness phases on seed and fruit quality of eggplant (*Solanum melongena* L.). Emirates Journal of Food and Agriculture. 2022;**34**(2):144-150

[76] Iwai T, Takahashi M, Oda K, Terada Y, Yoshida KT. Dynamic changes in the distribution of minerals in relation to phytic acid accumulation during rice seed development. Plant Physiology. 2012;**160**:2007-2014

[77] Lott JNA, Ockenden I, Raboy V, Batten GD. Phytic acid and phosphorus in crop grains, seeds and fruits. In: Reddy NR, editor. Food Phytate. Boca Raton, FL: CRC Press; 2002. pp. 7-24

[78] Raboy V. Approaches and challenges to engineering seed phytate and total phosphorus. Plant Science. 2009;**177**:281-296

[79] Loewus FA, Murthy PPN. Myo-inositol metabolism in plants. Plant Science. 2000;**150**:1-19

[80] White PJ, Broadley MR. Biofortification of crops with seven mineral elements often lacking in human diets–iron, zinc, copper, calcium, magnesium, selenium and iodine. New Phytologist. 2009;**182**:49-84

[81] Mandizvo T, Odindo AO. Seed mineral reserves and vigour of Bambara groundnut (*Vigna subterranea* L.) landraces differing in seed coat colour. Heliyon. 2019;**5**:e01635

[82] Mandizvo T, Odindo AO. Spectrophotometric quantification of phytic acid during embryogenesis in bambara groundnut (*Vigna subterranea* L.) through phosphomolybdenum complex formation. Emirates Journal of Food and Agriculture. 2020;**32**(11):778-785

[83] Zhawar VK, Kaur N, Gupta AK. Phytic acid and raffinose series oligosaccharides metabolism in developing chickpea seeds. Physiology and Molecular Biology of Plants. 2011;**17**:355-362

[84] Zhang WH, Zhou YC, Dibley KE, Tyerman SD, Furbank RT, Patrick JW. Nutrient loading of developing seeds. Functional Plant Biology. 2007;**34**:314-331

[85] Manavalan LP, Guttikonda SK, Tran LS, Nguyen HT. Physiological and molecular approaches to improve drought resistance in soybean. Plant Cell Physiology. 2009;**50**:1260-1276

[86] Joshi-Saha A, Reddy KS. Repeat length variation in the 5'UTR of myo-inositol monophosphatase gene is related to phytic acid content and contributes to drought tolerance in chickpea (*Cicer arietinum* L.). Journal of Experimental Botany. 2015;**66**:5683-5690

[87] Dhole VJ, Reddy KS. Association of phytic acid content with biotic stress tolerance in mungbean (*Vigna radiata* L. *Wilczek*). Phytoparasitica. 2016;**44**:261-267

Chapter 3

Nitrogen Assimilation and Translocation in Arabidopsis Seeds

Rowshon A. Begam and Michael Deyholos

Abstract

In plants, assimilated nitrogen travels mostly as amino acids. Amino acids travel from sources to sink tissues through cellular and organelle membranes such as plasma membrane, chloroplast membrane, mitochondrial membrane, and tonoplast membrane via facilitated or active transport. Membrane transporter proteins such as amino acid transporters mediate the transport. These transporters, as they facilitate the movement of amino acids through membranes, also regulate the distribution of amino nitrogen. Understanding the organ and tissue-specific distribution of amino acid transporters, their substrate affinity, and transport mechanism can help us understand the source-sink distribution of amino nitrogen in plants. With advancements in plant science research, we understand the amino acid distribution route in theory, but we have yet to identify many of the necessary amino acid transporters that enable this route. This chapter discusses the source-sink distribution of amino acids with a specific focus on seeds and lists the amino acid transporters in this route, characterized to date, in the model plant system, *Arabidopsis thaliana*.

Keywords: amino acids, amino nitrogen, seed nitrogen, amino acid distribution, seed nitrogen, seed storage nitrogen, Arabidopsis, amino acid transporter

1. Introduction

Plants take up organic nitrogen in the form of amino acids and peptides. However, the predominant forms of nitrogen taken up by plants are nitrate (NO_3^-) and ammonium (NH_4^+). Nitrogen, taken up as nitrate or ammonium, is assimilated into amino acids glutamine or glutamate [1]. When nitrogen is taken up as nitrate, it is reduced to ammonium before being assimilated into amino acids. Assimilated nitrogen travels as amino acids from the source to sink tissues.

The movement of amino acids from sources and sinks requires them to cross cellular and organelle membranes such as plasma membrane, chloroplast membrane, mitochondrial membrane, tonoplast membrane, and peroxisome membrane. Amino acids do not cross membranes through passive diffusion. When crossing a membrane, they require facilitated or active transport by membrane transporter proteins. These transporter proteins are known as amino acid transporters. Higher plants, or angiosperms, are complex organisms that have specialized organs and tissues for the uptake, transport, and storage of amino acids. This complex arrangement of sources and sink tissues requires a complex transport system involving amino acid export, import, antiport, or homeostasis mechanisms.

Additionally, amino acid transporters exhibit a preference for some amino acids over others. We explain this preference with terminologies such as "substrate affinity" and "substrate specificity." Amino acids of various charges and sizes such as acidic, basic, small neutral, or large neutral may require amino acid transporters with substrate affinity and substrate specificity for each kind to transport them across membranes. There are 20 common amino acids that are found in both plants and animals. These amino acids mainly function as building blocks for protein synthesis. In addition to the 20 common amino acids, plants produce several other nonproteinogenic amino acids for specialized functions.

The complexity of the amino acid transport process and the diversity in the types of amino acids require plants to have many amino acid transporters. **Table 1** shows the number of annotated amino acid transporters identified to date in selected monocot and dicot plant species that are fully sequenced. These numbers continue to evolve as more plant genomes are sequenced and the tools for phylogenetic analysis of gene sequences advance.

Arabidopsis thaliana, a dicotyledonous model plant species, has 85 to over 100 putative amino acid transporters [2, 3]. While these transporters facilitate the movement of amino acids through membranes, they also regulate the distribution of amino acids and, thus the partitioning of amino nitrogen. Understanding the distribution of amino acids in plants requires understanding the organ and tissue-specific distribution of amino acid transporters with their substrate affinity, substrate specificity, and transport mechanism. With advancements in plant biology research, we understand the amino acid distribution route in theory, although many of the necessary amino acid transporters in this route are obscure. This chapter focused on the amino nitrogen translocation and distribution in *Arabidopsis thaliana* with a specific focus on amino acid translocation to seeds and specifies the amino acid transporters in this route that are functionally characterized to date.

Amino acids have diverse functions in plants beyond their role in protein synthesis, such as signaling molecules, osmolytes, antioxidants, precursors for secondary metabolites, and regulators of gene expression. This chapter will discuss translocation, storage, and distribution of amino acids in seeds regardless of the specific functions these amino acids may conduct. The number and classification of amino acid transporters in plants continue to evolve as genome sequencing and annotation technologies advance, and as new research is conducted in this area.

Plant species	Total number of annotated amino acid transporters	Reference(s)
Arabidopsis thaliana	85–100	[2, 3]
Wheat (*Triticum aestivum*)	283	[4]
Rice (*Oryza sativa* L.)	85	[5]
Soybean (*Glycine max*)	189	[6]
Tomato (*Solanum lycopersicum*)	88	[7]
Potato (*Solanum tuberosum* L)	72	[8]

Table 1.
Annotated amino acid transporters in Arabidopsis thaliana *and selected crop species.*

The scope of this chapter is limited to the functionally characterized or annotated amino acid transporters to date.

2. Source and sink tissues for amino acids in *Arabidopsis thaliana*

2.1 Amino acid sources and sinks in the whole plant

It is hard to distinguish organs or tissues as the sole source or sink of amino acids. Most tissues participate in both import and export of amino acids throughout their developmental stages. We call the tissues or organs that play as net exporters of amino acids "source tissues" and the tissues or organs that play as net importers of amino acids "sink tissues." In general, photosynthetically active green tissues are sources of amino acids. Amino acids are produced primarily in photosynthetically active leaves and, to a lesser extent, in other green tissues such as stems and flowers, and in roots. Senescing leaves do not produce amino acids, but as they die and decay, they recycle the cellular components and become a source of amino acids. Once amino acids are synthesized or recycled in the source tissues, they are transported to various parts of the plant where they are needed. Sink tissues in plants include actively growing tissues such as shoot and root apex, developing seeds, and fruits. These tissues have a high demand for amino acids for protein synthesis and other metabolic processes, seeds being the final sink that stores amino acids as a storage protein.

2.2 Amino acid sources for seeds

At the postfertilization stage in Arabidopsis, green carpel cells in the fruit develop distinctive features with well-defined stomata in the epidermal cells for gas exchange and three layers of mesophyll tissue with photosynthetic capacity [9]. Due to the profusion of open stomata in green fruits, the transpiration pull of xylem sap can deliver amino acids from roots, green leaves, or senescing leaves to the green carpel cells. Fruit carpel cells, therefore, may play an important role in seed nutrition. However, as the stream of the xylem sap is stronger toward the leaves due to the higher rate of transpiration in the leaves, amino acids loaded in the xylem from the root are translocated predominantly to the leaves, where they are temporarily stored or metabolized before being transported to the seeds [10]. Green leaves and roots serve as sources of amino acids for seeds during the early reproductive stages. However, most species, including Arabidopsis, accumulates seed storage compounds concomitantly with the acquisition of dormancy and desiccation tolerance [11]. During this stage, recycled amino acids in leaves derived from photorespiration and leaf senescence feed the reproductive sink tissues [12]. During seed maturation, up to 80 per cent of seed amino acids may come from leaves, especially from the senescing leaves [13, 14].

3. Amino acid transporters in Arabidopsis siliques and seeds

The most recent report suggests that there are more than 100 annotated amino acid transporters in *Arabidopsis thaliana* (**Table 1**), although the number varies in other published reports [3]. Based on expression analysis, 22 of these amino acid transporters are expressed in siliques and seeds (**Table 2**). These transporters mostly belong to the cationic amino acid transporter (CAT) family, and usually multiple

Name(s)	Gene ID/ Locus	Transporter family	Tissue expression	Subcellular localization	Possible transport type
AAP1/NAT2	AT1G58360	AAP	Siliques	PM	Im
AAP2	AT5G09220	AAP	Siliques	–	Im
AAP5	AT1G44100	AAP	Siliques	–	Im
AAP8	AT1G10010	AAP	Silique and seed	PM	Im
BAT1/GABP	AT2G01170	ACT	Silique	MM	Ex/Im
CAT1	AT4G21120	CAT	Silique	PM	Im
CAT2	AT1G58030	CAT	Silique	TM	–
CAT3	AT5G36940	CAT	Silique	ER	Im
CAT4	AT3G03720	CAT	Silique	TM	–
CAT5	AT2G34960	CAT	Silique	PM	Im
CAT6	AT5G04770	CAT	Silique and seed	PM	Im
CAT8	AT1G17120	CAT	Silique	TM and PM	Im
LAT4/PUT2/ PAR1	AT1G31830	PHS/LAT	Silique and seed	GA	Ex/Im
LAT5	AT3G19553	PHS/LAT	Silique	ER	Ex/Im
LHT1	AT5G40780	LHT	Silique	PM	Im
UMAMIT11	At2g40900	UMAMIT	Silique	PM	Ex/Im
UMAMIT14	At2g39510	UMAMIT	Silique	PM	Ex/Im
UMAMIT18/ SIAR1	At1g44800	UMAMIT	Silique	PM	Ex/Im
UMAMIT24	At1g25270	UMAMIT	Seed (coat)	TM	Ex/Im
UMAMIT25	At1g09380	UMAMIT	Seed (endosperm)	PM	Ex/Im
UMAMIT28	At1g01070	UMAMIT	Silique (mature)	PM	Ex/Im
UMAMIT29	At4g01430	UMAMIT	Silique (young)	PM	Ex/Im

AAP, amino acid permease; LHT, lysine histidine transporter; CAT, cationic amino acid transporter; ACT, amino acid choline transporter; PHS, polyamine H + —symporters; LAT, L-type amino acid transporter; UMAMIT, usually multiple amino acids move in and out transporter; PM, plasma membrane; ER, endoplasmic reticulum; GA, Golgi apparatus; MM, mitochondrion membrane; TM, tonoplast membrane; –, unknown; Ex, export; Im, import.
Source: [3, 15–21] and references therein.

Table 2.
Annotated amino acid transporters expressed in Arabidopsis siliques and seeds. The transporters listed in this table may also express in other organs and tissues.

amino acids move in and out transporter (UMAMIT) family, followed by the amino acid permease (AAP) family. Acquisition of amino acids in seeds can be influenced by any amino acid transporter expressed in the source and sink tissues or along the vascular transport path. However, the CAT, UMAMIT, and AAP families, with notable members expressed in seeds and siliques, appear to have significant importance in amino acid transport to seeds. There was a gap in understanding the amino acid export process in plants until the recent identification of the UMAMIT family. Members of this family are capable of both importing and exporting amino acids and are localized in both plasma and organelle membranes [22–24].

4. Amino acid mobilization to seeds

Mobilization of amino acids from source tissues to sink tissues involves two steps: intracellular movement through organelle membranes and cell-to-cell or long-distance movement through plasma membrane.

4.1 Intracellular movement of amino acids through organelle membrane

4.1.1 Chloroplast

Primary assimilation of inorganic nitrogen and reassimilation of recycled nitrogen into amino acids occur through a series of chemical reactions taking place in both cytosol and plastid or chloroplast [25, 26]. During primary assimilation, NO_3^- taken up by plants is transported to plastids or chloroplast where it is reduced to NH_4^+ and subsequently assimilated into glutamine via glutamate. Reassimilation of recycled NH_4^+, derived from photorespiration or protein hydrolysis, also occurs in the chloroplast [25, 26]. During photorespiration in C_3 plants, such as Arabidopsis, the 2-carbon compound produced through the oxygenase activity of Rubisco needs to be converted to a 3-carbon compound so that the photosynthetically fixed carbon can be rescued and fed back into the Calvin cycle (**Figure 1**). This occurs through a complex series of biochemical reactions taking place in the chloroplast, peroxisome, and mitochondria. The 2-carbon compound (2-phosphoglycolate) finally leads to the formation of glycine and a 3-carbon compound (3-phosphoglycerate) [27, 28]. During this process, both nitrogen and carbon are released in the forms of NH_3 and CO_2 in the mitochondria. This photorespiratory NH_3 is recaptured by the GS2/Fdx-GOGAT pathway in the chloroplast. Thus, chloroplasts/plastids are vital organelles for nitrogen assimilation and remobilization [29, 30]. Nitrogen, stored in the forms of storage proteins, peptides, or amino acids, is mobilized primarily in the form of amino acids. Most amino acids required for protein synthesis are produced in the chloroplast/plastid. Plants follow a basipetal growth pattern, where developing tissues depend on mature tissues for the supply of amino acids for protein synthesis. Amino acids, synthesized in the chloroplast/plastid in mature cells, are subject to both intracellular and long-distance transport. Amino acids biosynthesized in the chloroplast/plastid cannot cross the inner and outer membrane without a membrane transporter or a channel protein. Plants thus need both amino acid export- and import systems in the inner and outer membranes of chloroplast. In the outer membrane of the chloroplast, Outer Envelope Protein 16 and 24 (OEP16 & 24) facilitate amino acid transport. In Arabidopsis, several OEP16- and OEP24-family genes have been identified that may mediate amino acid transport through the outer membrane of the chloroplast/plastid [30–32]. Microarray analysis and in silico subcellular localization analysis have identified putative amino acid transporters that may be localized in the inner chloroplast membrane [33, 34]. A Glutamate/Malate antiporter (DiT2 coupled with DiT1) in the inner chloroplast membrane mediates glutamate export from the stroma in exchange for malate (**Figure 1**) [35–37]. Not many amino acid transporters, with a net export capacity, in the inner chloroplast membrane are known to date.

4.1.2 Mitochondria

Plant mitochondria are also important in intracellular nitrogen metabolism, including the synthesis and catabolism of amino acids [38, 39]. Mitochondria,

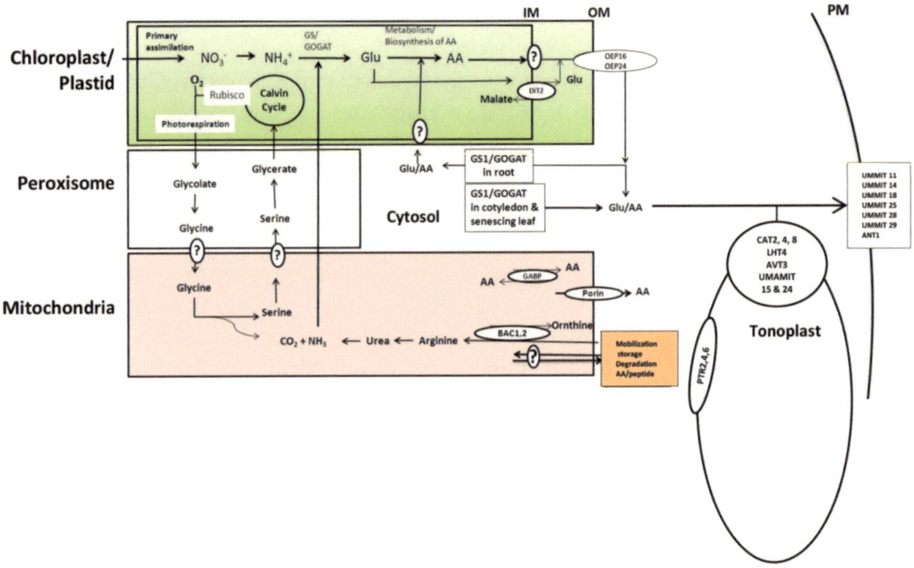

Figure 1.

A simplified model of intracellular amino acid transport shows the movement of amino acids in- and out of membrane-bound organelles in a plant cell. Based on the available information to date, amino acid transporters yet to be identified have been indicated with a '?' mark within a circle. Nitrogen metabolism in chloroplast and plastid has been shown together in the same organelle. AA, amino acid; GS, glutamine synthetase; GOGAT, glutamate synthase; IM, inner membrane; OM, outer membrane; PM, plasma membrane. Primary assimilation of nitrate (NO_3^-) takes place in the chloroplast or plastid. Amino acids derived from the primary assimilation of nitrogen that occurs through cytosolic GS/GOGAT are imported into the chloroplast or plastid for the biosynthesis of other amino acids. All amino acids synthesized in the chloroplast or plastid are exported into the cytosol for cellular use or translocation. Channel proteins OEP16 & 24 in the outer envelope mediate amino acid transport. In the inner membrane, glutamate/malate antiporter (DiT2) mediates glutamate export. No other amino acid transporters are known to mediate amino acid import or export through the inner membrane. Amino acids derived from protein hydrolysis are catabolized in the mitochondria. Reserve nitrogen enters uni-directionally into the mitochondria to be catabolized during seed germination. During photorespiration, glycine produced in the peroxisome enters mitochondria where it is converted to serine and exported back to peroxisome. In the outer envelope, porins mediate amino acid transport. In the inner membrane arginine/ornithine antiporter (BAC1, 2) and a bi-directional transporter (GABP) are known so far. In the tonoplast, temporary storage of amino acids and subsequent release requires amino acid transporters with both export and import capacity. Arabidopsis CAT2, 4, & 8, LHT4, AVT3, UMAMIT15 & 24 are localized in the tonoplast membrane and may function as vacuolar amino acid transporters. Three peptide transporters (PTR2, 4, 6) are also localized in the tonoplast membrane with unknown transport direction.

together with chloroplasts and peroxisomes, manage both photorespiration and photosynthesis along with many other metabolic pathways [40]. Mitochondria also play an important role in the remobilization of storage nitrogen during seed germination [41]. Characterization of transporter proteins involved in the transport of nitrogen compounds in the inner and outer mitochondrial membrane will contribute to a better understanding of the role of mitochondria in nitrogen metabolism and distribution. More than 50 genes in Arabidopsis have been annotated to encode mitochondrial carrier proteins, and several proteins have been speculated to encode amino acid transporters and localized in the inner and outer envelope of mitochondria [41–43]. In the inner mitochondrial membrane, two basic amino acid carriers, BAC1 and BAC2, have been experimentally shown to be involved in amino acid transport [44, 45]. These carrier proteins mediate arginine/ornithine (or Citrulline) antiport in the inner mitochondrial membrane (**Figure 1**). To exchange glycine and serine with the peroxisome during photorespiration, there might be an exchanger or a glycine/serine antiporter in the inner mitochondrial membrane that is yet to be identified. A report has shown that the GABP (also known as BAT1) in Arabidopsis is localized in the mitochondrial membrane [46]. The transporter mediates both the import and export of amino acids, suggesting that it might be a bi-directional facilitator [21]. The outer membrane of mitochondria is permeable to solutes up to a size of 4–5 kDa through porins [47]. The average size of an amino acid is much smaller than 5 kDa. The mitochondrial porins in the outer membrane, also called VDAC (Voltage Dependent Anion Channels), that were characterized as relatively nonspecific general diffusion pores may mediate amino acid transport through the outer envelope [48]. Movement of amino acids through porins in the outer envelope of mitochondria has yet to be studied in plants.

4.1.3 Tonoplast

Amino acids are temporarily stored in the vacuole and subsequently released into the cytoplasm. This process requires amino acid transporters in the tonoplast membrane with export and import capacity. The Arabidopsis CAT2, 4, & 8, LHT4, AVT3, UMAMIT15 & 24 are localized to the tonoplast membrane and may function as vacuolar amino acid transporters (reviewed in [3, 15]). Arabidopsis PTR2, 4, & 6, members of the PTR/NRT1 family, were shown to be localized in the tonoplast membrane and are candidates to mediate peptide transport in and out of the tonoplast [49].

4.2 Amino acid translocation from source tissues to seeds

4.2.1 Loading amino acids from leaf mesophyll cells into the phloem minor vein

Regardless of the source, amino acids travel to seeds via both xylem and phloem but are delivered into seed sink tissues via phloem minor veins [50, 51]. Loading amino acids from leaf mesophyll cells into the phloem minor vein may occur both symplastically and apoplastically. While symplastic loading via the plasmodesmata can be rate-limiting [52], it is improbable in some species since the solute concentration in sieve elements and companion cells in the phloem can be much greater than those in the surrounding source cells. High solute concentration enables the hydrostatic pressure in the phloem that drives long-distance transport of solutes [52–55]. Thus, in many species including Arabidopsis, loading assimilates into the phloem occurs apoplastically [52, 56]. In the apoplastic loading, amino acids are exported from mesophyll cells into the apoplasm, followed by active uptake into the sieve element-companion cell complex of the phloem [52, 57–60]. While amino acid exporters in leaf mesophyll cell plasma membrane are obscure, published reports suggest that Arabidopsis AAP2, 5, & 8, CAT6 & 9, ProT1, and LAT5 are either expressed in the phloem or demonstrated to have a role in phloem loading [16, 19, 51, 61–66]. The LAT4 is expressed in green carpel cells in the silique with a possible role in mobilizing amino acids from these tissues toward seeds [20]. The recently identified UMAMIT facilitators (UMAMIT 14, 18, 28, &29) that are expressed in phloem may also have a role in phloem loading and unloading [15, 22]. **Figure 2** shows the possible role of the amino acid transporters characterized to date.

4.2.2 Phloem-xylem-phloem exchange

Amino acids, loaded into the phloem from source tissues in the leaf, may undergo transfer from phloem to xylem for upward translocation. The importance

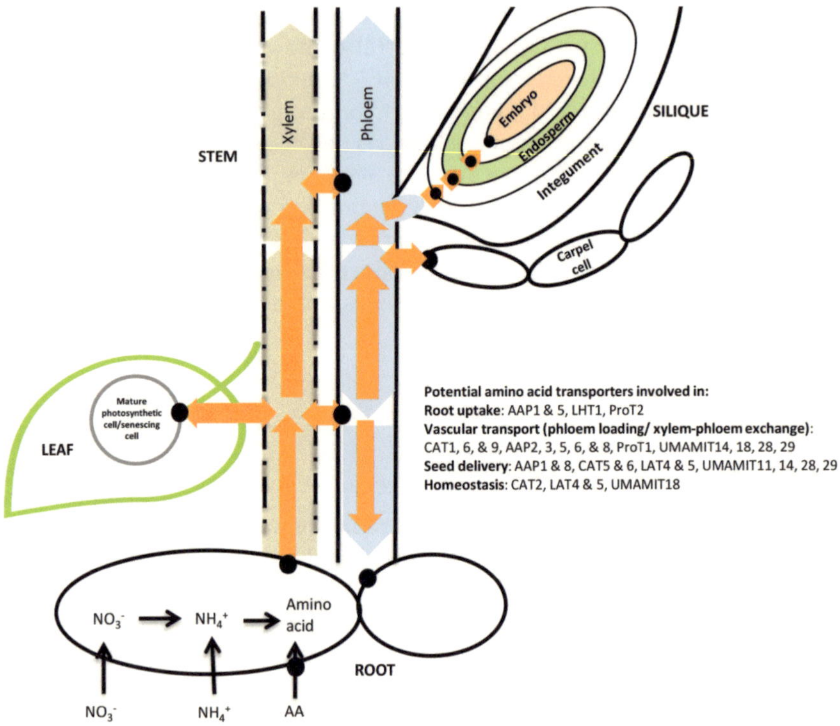

Figure 2.
A simplified model of amino acid transport from source to sink tissues in Arabidopsis thaliana*. The dark circles show the positions of one or more amino acid transporters involved in the route with import, export, or bi-directional facilitator capacity. This figure represents the plasma membrane crossing between symplasm and apoplasm for amino acid translocation from source tissues to the seed embryo. The orange arrows indicate directions of amino acid transport. Transporters for other forms of nitrogen are not shown in this figure.*

of this phloem-xylem exchange for amino acid distribution within plants has been demonstrated in several physiological studies [67–69]. However, at the end of the long-distance transport through the xylem, amino acids are loaded back to the phloem because in Arabidopsis, amino acids are delivered to seeds via the phloem [52]. Exchanges of amino acids from phloem to xylem or xylem to phloem are an exchange between symplasm and apoplasm, and thus require amino acid transporters in the plasma membrane of phloem companion cells with a net export or import capacity. In Arabidopsis, CAT1, 6, & 9, AAP2, 3, 5, 6, & 8, ProT1, UMAMIT14, 18, 28, 29 are either expressed in the vascular tissues or demonstrated to have a role in phloem-xylem-phloem exchange of amino acids. For example, the expression of the AAP2 [51, 70] along the vascular transport strand in the stem indicated its involvement in active exchange of amino acids between the xylem and phloem. The AAP2 is expressed in the phloem in the stem, and in funiculi in the silique [70]. It is an import transporter and, therefore, plays a potential role in xylem-to-phloem loading and delivering amino acids to the seed [51]. The AAP6 is expressed in the xylem parenchyma, mediating amino acid import in heterologous system [17]. An in-planta study showed that knocking out the AAP6 reduces total amino acid concentration in the phloem suggesting an indirect role in loading amino acids into the phloem [71].

4.2.3 Loading into the seeds

In Arabidopsis siliques, phloem terminates at the funiculus, and the seed outer integument cells work as a symplastic extension of the funicular phloem [72]. However, transferring amino acids from the outer integument to the inner integument, from the inner integument to the embryo, and from the embryo to the endosperm, may be apoplastic. The seed embryo is separated from the mother plant phloem by three apoplastic borders that require amino acid transporters with export and import capacity at each border to transfer amino acids from the funicular phloem into the embryo [72]. The recently identified UMAMIT 11, 14, 28, & 29 are expressed in siliques in tissues adjacent to phloem from which amino acids are usually exported. Knocking out these genes results in accumulating free amino acids in fruits and producing smaller seeds. These plasma membrane-localized facilitators are a good candidate to facilitate amino acid export and import in seeds [3, 15, 22]. The UMAMIT24 is expressed in the chalazal seed coat but is localized in the tonoplast membrane [23]. Its direct role is likely in the intracellular movement of amino acids. The UMAMIT25 is a plasma membrane-localized transporter expressed in the seed endosperm suggesting a possible role in amino acid transfer between embryo and endosperm [23]. The AAP8 plays a role in importing amino acids into the endosperm and supplying the developing embryo with amino acids during early embryogenesis [73]. The Arabidopsis AAP1 is expressed in the developing embryo during embryo morphogenesis and early maturation in the filial tissues and plays a role in importing amino acids into the embryo [70, 74–76]. Knocking out the AAP1 gene caused amino acids to accumulate in the seed coat/endosperm [76]. The CAT6 is expressed in the seed with a possible role in amino acid distribution within the seed [64]. The LAT5 is expressed in phloem and in siliques, and it possibly has a bi-directional amino acid transport capacity. However, knocking out this transporter caused increased nitrogen content in seeds [19]. It probably plays a role in amino acid homeostasis rather than importing amino acids into seeds. In Arabidopsis, the endosperm degenerates during early seed maturation, and the embryo becomes the final sink of storage protein, while in other species, the endosperm serves as the source of nutrients during seed germination [77]. Regardless, the transfer of amino acids from the endosperm into the embryo is important in terms of seed protein content and yield achievement. Identification of amino acid transporter with net export capacity in the inner and outer integument cells will allow a clearer understanding of amino acid distribution mechanism during seed maturity.

5. Conclusion

Crop improvement research strives to understand the amino nitrogen (i.e., amino acids) distribution process in plants primarily to improve crops' nitrogen use efficiency and grain nutritional quality by improving the protein content in seeds. With advancements in plant science research, we understand the amino acid distribution route in theory but many of the necessary amino acid transporters that enable this route are still unknown. Identification and characterization of amino acid transporters in plants have advanced significantly in the recent decade, although many more are still unknown. In Arabidopsis, there are at least 100 annotated amino acid transporters, with more than 20 expressed in the seeds and siliques alone. This suggests a robust and complex process regulating plant amino acid distribution and protein storage in seeds. The process is further nuanced at various developmental stages or as plants respond to biotic and abiotic factors. We need to understand the organ and tissue-specific distribution of all amino acid transporters, their substrate affinity, and transport mechanism to understand the source-sink distribution of amino acids in plants fully.

Acknowledgements

We are thankful to the reviewers, the editors, and the editorial office for their technical and scientific support in compiling this chapter.

Author details

Rowshon A. Begam[1*] and Michael Deyholos[2]

1 University of Alberta, Edmonton, Alberta, Canada

2 University of British Columbia, Kelowna, Canada

*Address all correspondence to: rowshon.begam@gov.ab.ca

References

[1] Tempest DW, Meers JL, Brown CM. Synthesis of glutamate in Aerobacter aerogenes by a hitherto unknown route. Biochemical Journal. 1970;**117**(2):405-407

[2] Gao C, Wang Y, Jiang L. Plant membrane-bound transporters for amino acids: Roles in nutrient homeostasis, signalling, and stress tolerance. Journal of Experimental Botany. 2016;**67**(14):4401-4419

[3] Dhatterwal P, Mehrotra S, Miller A, Mehrotra R. Promoter profiling of Arabidopsis amino acid transporters: Clues for improving crops. Plant Molecular Biology. 2021;**107**(6):451-575

[4] Wan Y, King R, Mitchell RAC, Hassani-Pak K, Hawkesford MJ. Spatiotemporal expression patterns of wheat amino acid transporters reveal their putative roles in nitrogen transport and responses to abiotic stress. Scientific Reports. 2017;7:5460

[5] Zhao H, Ma H, Yu L, Wang X, Zhao J. Genome-wide survey and expression analysis of amino acid transporter gene family in rice (Oryza sativa L.). PLoS One. 2012;7(11):e49210

[6] Cheng L, Yuan H-Y, Ren R, Zhao S-Q, Han Y-P, Zhou Q-Y, et al. Genome-wide identification, classification, and expression analysis of amino acid transporter gene family in glycine max. Frontiers in Plant Science. 2016;**7**:1-15

[7] Alzahrani FO. Genome wide analysis of amino acid transporter superfamily in Solanum lycopersicum. Plants. 2021;**10**(2):289

[8] Ma H, Cao X, Shi S, Li S, Gao J, Ma Y, et al. Genome-wide survey and expression analysis of the amino acid transporter superfamily in potato (Solanum tuberosum L.). Plant Physiology and Biochemistry. 2016;**107**:164-177

[9] Dinneny JR, Yanofsky MF. Drawing lines and Borders: How the dehiscent fruit of Arabidopsis is patterned. BioEssays. 2004;**27**(1):42-49

[10] Tegeder M, Rentsch D. Uptake and partitioning of amino acids and peptides. Molecular Plant. 2010;**3**(6):997-1011

[11] Goldberg RB, Beals TP, Sanders PM. Anther development: Basic principles and practical applications. Plant Cell. 1993;**5**(10):1217-1229

[12] Peoples MB, Dalling MJ. The interplay between proteolysis and amino acid metabolism during senescence and nitrogen reallocation. In: Nooden LD, Leopold AC, editors. Senescence and Aging in Plants. 1988. pp. 181-217. Available from: http://hdl.handle.net/102.100.100/263015?index=1

[13] Yuan ZY, Li LH, Han XG, Huang JH, Jiang GM, Wan SQ, et al. Nitrogen resorption from senescing leaves in 28 plant species in a semi-arid region of northern China. Journal of Arid Environments. 2005;**63**(1):191-202

[14] Aerts R, Chapin FS III. The mineral nutrition of wild plants revisited: A Re-evaluation of processes and patterns. Advances in Ecological Research. 1999;**30**:1-67

[15] Yao X, Nie J, Bai R, Sui X. Amino acid transporters in plants: Identification and function. Plants. 2020;**9**(8):972

[16] Su Y-H, Frommer WB, Ludewig U. Molecular and functional characterization

of a family of amino acid transporters from Arabidopsis. Plant Physiology. 2004;**136**(2):3104-3113

[17] Okumoto S, Schmidt R, Tegeder M, Fischer WN, Rentsch D, Frommer WB, et al. High affinity amino acid transporters specifically expressed in xylem parenchyma and developing seeds of Arabidopsis. Journal of Biological Chemistry. 2002;**277**(47):45338-45346

[18] Yang H, Krebs M, Stierhof Y-D, Ludewig U. Characterization of the putative amino acid transporter genes AtCAT2, 3 &4: The tonoplast localized AtCAT2 regulates soluble leaf amino acids. Journal of Plant Physiology. 2014;**171**(8):594-601

[19] Begam RA, D'Entremont J, Good A. The Arabidopsis L-type amino acid transporter 5 (LAT5/PUT5) is expressed in the phloem and alters seed nitrogen content when knocked out. Plants. 2020;**9**(11):1519

[20] Begam RA, Good AG. The Arabidopsis paraquat resistant1 mutant accumulates leucine upon dark treatment. Botany. 2017;**95**(7):751-761

[21] Dündar E, Bush DR. BAT1, a bidirectional amino acid transporter in Arabidopsis. Planta. 2009;**229**:1047-1056

[22] Müller B, Fastner A, Karmann J, Mansch V, Hoffmann T, Schwab W, et al. Amino acid export in developing Arabidopsis seeds depends on UMAMIT facilitators. Current Biology. 2015;**25**(23):3126-3131

[23] Besnard J, Zhao C, Avice J-C, Vitha S, Hyodo A, Pilot G, et al. Arabidopsis UMAMIT24 and 25 are amino acid exporters involved in seed loading. Journal of Experimental Botany. 2018;**69**(21):5221-5232

[24] Zhao C, Pratelli R, Yu S, Shelley B, Collakova E, Pilot G. Detailed characterization of the UMAMIT proteins provides insight into their evolution, amino acid transport properties, and role in the plant. Journal of Experimental Botany. 2021;**72**(18):6400-6417

[25] Liu X, Hu B, Chu C. Nitrogen assimilation in plants: Current status and future prospects. Journal of Genetics and Genomics. 2022;**49**(5):390-404

[26] Masclaux-Daubresse C, Daniel-Vedele F, Dechorgnat J, Chardon F, Gaufichon L, Suzuki A. Nitrogen uptake, assimilation and remobilization in plants: Challenges for sustainable and productive agriculture. Annals of Botany. 2010;**105**(7):1141-1157

[27] Maurino VG, Peterhansel C. Photorespiration: Current status and approaches for metabolic engineering. Current Opinion in Plant Biology. 2010;**13**(3):248-255

[28] Peterhansel C, Maurino VG. Photorespiration redesigned. Plant Physiology. 2011;**155**(1):49-55

[29] Lopez-Juez E, Pyke KA. Plastids unleashed: Their development and their integration in plant development. The International Journal of Developmental Biology. 2005;**49**:557-577

[30] Pudelski B, Kraus S, Soll J, Philippar K. The plant PRAT proteins–preprotein and amino acid transport in mitochondria and chloroplasts. Plant Biology. 2010;**12**(s1):42-55

[31] Duy D, Soll J, Philippar K. Solute channels of the outer membrane: From bacteria to chloroplasts. Biological Chemistry. 2007;**388**(9):879-889

[32] Pottosin I, Shabala S. Transport across chloroplast membranes: Optimizing photosynthesis for adverse

environmental conditions. Molecular Plant. 2016;**9**(3):356-370

[33] Chloroplast Function Database II [Online]. RIKEN Plant Science Center; 2023. Available from: http://rarge-v2.psc.riken.jp/chloroplast/ [Accessed June 6, 2023]

[34] Koo AJK, Ohlrogge JB. The predicted candidates of Arabidopsis plastid inner envelope membrane proteins and their expression profiles. Plant Physiology. 2002;**130**(2):823-836

[35] Linka M, Weber APM. Shuffling ammonia between mitochondria and plastids during photorespiration. Trends in Plant Science. 2005;**10**(10):461-465

[36] Philippar K, Soll J. Intracellular transport: Solute transport in chloroplasts, mitochondria, peroxisomes and vacuoles, and between organelles. In: Yeo A, Flowers T, editors. Plant Solute Transport. Oxford, UK: Blackwell Publishing Ltd; 2007

[37] Weber APM, Fischer K. Making the connections–the crucial role of metabolite transporters at the Interface between chloroplast and cytosol. FEBS Letters. 2007;**581**(12):2215-2222

[38] Mackenzie S, McIntosh L. Higher plant mitochondria. The Plant Cell. 1999;**11**(4):571-585

[39] Bowsher CG, Tobin AK. Compartmentation of metabolism within mitochondria and plastids. Journal of Experimental Botany. 2001;**52**(356):513-527

[40] Raghavendra AS, Padmasree K. Beneficial interactions of mitochondrial metabolism with photosynthetic carbon assimilation. Trends in Plant Science. 2003;**8**(11):546-553

[41] Picault N, Hodges M, Palmieri L, Palmieri F. The growing family of mitochondrial carriers in Arabidopsis. Trends in Plant Science. 2004;**9**(3):138-146

[42] Millar AH, Heazlewood JL. Genomic and proteomic analysis of mitochondrial carrier proteins in Arabidopsis. Plant Physiology. 2003;**131**(2):443-453

[43] Murcha M, Elhafez D, Lister R, Tonti-Filippini J, Baumgartner M, Philippar K, et al. Characterization of the Preprotein and amino acid transporter gene family in Arabidopsis. Plant Physiology. 2007;**143**(1):199-212

[44] Catoni E, Desimone M, Hilpert M, Wipf D, Kunze R, Schneider A, et al. Expression pattern of a nuclear encoded mitochondrial arginine-ornithine translocator gene from Arabidopsis. BMC Plant Biology. 2003;**3**(1):1-10

[45] Hoyos ME, Palmieri L, Wertin T, Arrigoni R, Polacco JC, Palmieri F. I dentification of a mitochondrial transporter for basic amino acids in Arabidopsis thaliana by functional reconstitution into liposomes and complementation in yeast. The Plant Journal. 2003;**33**(6):1027-1035

[46] Michaeli S, Fait A, Lagor K, Nunes-Nesi A, Grillich N, Yellin A, et al. A mitochondrial GABA permease connects the GABA shunt and the TCA cycle, and is essential for normal carbon metabolism. The Plant Journal. 2011;**67**:485-498

[47] Benz R. Permeation of hydrophilic solutes through mitochondrial outer membranes: Review on mitochondrial porins. Biochimica et Biophysica Acta (BBA)-Reviews on Biomembranes. 1994;**1197**(2):167-196

[48] Mannella CA. Minireview: On the structure and gating mechanism of the mitochondrial channel, VDAC. Journal

of Bioenergetics and Biomembranes. 1997;**29**:525-531

[49] Weichert A, Brinkmann C, Komarova NY, Dietrich D, Thor K, Meier S, et al. AtPTR4 and AtPTR6 are differentially expressed, tonoplast-localized members of the peptide transporter/nitrate transporter 1 (PTR/NRT1) family. Planta. 2012;**235**:311-323

[50] Rentsch D, Schmidt S, Tegeder M. Transporters for uptake and allocation of organic nitrogen compounds in plants. FEBS Letters. 2007;**581**(12):2281-2289

[51] Zhang L, Tan Q, Lee R, Trethewy A, Lee Y-H, Tegeder M. Altered xylem-phloem transfer of amino acids affects metabolism and leads to increased seed yield and oil content in Arabidopsis. The Plant Cell. 2010;**22**(11):3603-3620

[52] Lalonde S, Tegeder M, Throne-Holst M, Frommer WB, Patrick JW. Phloem loading and unloading of sugars and amino acids. Plant, Cell & Environment. 2003;**26**(1):37-56

[53] Geiger DR, Giaquinta RT, Sovonick SA, Fellows RJ. Solute distribution in sugar beet leaves in relation to phloem loading and translocation. Plant Physiology. 1973;**52**(6):585-589

[54] Turgeon R. Plasmodesmata and solute exchange in the phloem. Australian Journal of Plant Physiology. 2000;**27**(6):521-529

[55] Van Bel AJE. Strategies of phloem loading. Annual Review of Plant Physiology and Plant Molecular Biology. 1993;**44**:253-281

[56] Turgeon R, Wolf S. Phloem transport: Cellular pathways and molecular trafficking. Annual Review of Plant Biology. 2009;**60**:207-221

[57] Williams LE, Miller AJ. Transporters responsible for the uptake and partitioning of nitrogenous solutes. Annual Review of Plant Physiology and Plant Molecular Biology. 2001;**52**:659-688

[58] Winter H, Lohaus G, Heldt HW. Phloem transport of amino acids in relation to their cytosolic levels in barley leaves. Plant Physiology. 1992;**99**(3):996-1004

[59] Ortiz-Lopez A, Chang H-C, Bush DR. Amino acid transporters in plants. Biochimica et Biophysica Acta (BBA)-Biomembranes. 2000;**1465**(1-2):275-280

[60] Delrot S, Rochat C, Tegeder M, Frommer W. Amino Acid Transport. In: Lea PJ, Morot-Gaudry JF, editors. Plant Nitrogen. Berlin, Heidelberg: Springer; 2001. pp. 213-235

[61] Brady SM, Orlando DA, Lee J-Y, Wang JY, Koch J, Dinneny JR, et al. A high-resolution root spatiotemporal map reveals dominant expression patterns. Science. 2007;**318**(5851):801-806

[62] Fischer W-N, Kwart M, Hummel S, Frommer WB. Substrate specificity and expression profile of amino acid transporters (AAPs) in Arabidopsis. Journal of Biological Chemisty. 1995;**270**(27):16315-16320

[63] Santiago JP, Tegeder M. Connecting source with sink: The role of Arabidopsis AAP8 in phloem loading of amino acids. Plant Physiology. 2016;**171**(1):508-521

[64] Hammes UZ, Nielsen E, Honaas LA, Taylor CG, Schachtman DP. AtCAT6, a sink-tissue-localized transporter for essential amino acids in Arabidopsis. Plant Journal. 2006;**48**(3):414-426

[65] Grallath S, Weimar T, Meyer A, Gumy C, Suter-Grotemeyer M, Neuhaus

J-M, et al. The AtProT family. Compatible solute transporters with similar substrate specificity but differential expression patterns. Plant Physiology. 2005;**137**(1):117-126

[66] Rentsch D, Hirner B, Schmelzer E, Frommer WB. Salt stress-induced proline transporters and salt stress-repressed broad specificity amino acid permeases identified by suppression of a yeast amino acid permease-targeting mutant. Plant Cell. 1996;**8**(8):1437-1446

[67] Pate JS, Sharkey PJ, Lewis OA. Xylem to phloem transfer of solutes in fruiting shoots of legumes, studied by a phloem bleeding technique. Planta. 1975;**122**(1):11-26

[68] Schobert C, Komor E. Amino acid uptake by Ricinus Communis roots: Characterization and physiological significance. Plant, Cell & Environment. 1987;**10**(6):493-500

[69] Atkins C. Biochemical aspects of assimilate transfers along the phloem path: N-solutes in Lupins. Australian Journal of Plant Physiology. 2000;**27**(6):531-537

[70] Hirner B, Fischer WN, Rentsch D, Kwart M, Frommer WB. Developmental control of H+/amino acid permease gene expression during seed development of Arabidopsis. The Plant Journal. 1998;**14**(5):535-544

[71] Hunt E, Gattolin S, Newbury H, Bale J, Tseng H, Barrett D, et al. A mutation in amino acid permease AAP6 reduces the amino acid content of the Arabidopsis sieve elements but leaves aphid herbivores unaffected. Journal of Experimental Botony. 2010;**61**(1):55-64

[72] Baud S, Dubreucq B, Miquel M, Rochat C, Lepiniec L. Storage Reserve Accumulation in Arabidopsis: Metabolic and Developmental Control of Seed Filling. Vol. 2008. Rockville, USA: American Society of Plant Biologists; 2008

[73] Schmidt R, Stransky H, Koch W. The amino acid permease AAP8 is important for early seed development in Arabidopsis thaliana. Planta. 2007;**226**(4):805-813

[74] Boorer KJ, Frommer WB, Bush DR, Kreman M, Loo DDF, Wright EM. Kinetics and specificity of a H+/amino acid transporter from Arabidopsis thaliana. Journal of Biological Chemistry. 1996;**271**(4):2213-2220

[75] Fischer W-N, Loo DDF, Koch W, Ludewig U, Boorer KJ, Tegeder M, et al. Low and high affinity amino acid H+-cotransporters for cellular import of neutral and charged amino acids. The Plant Journal. 2002;**29**(6):717-731

[76] Sanders A, Collier R, Trethewy A, Gould G, Sieker R, Tegeder M. AAP1 regulates import of amino acids into developing Arabidopsis embryos. The Plant Journal. 2009;**59**(4):540-552

[77] Hill LM, Morley-Smith ER, Rawsthorne S. Metabolism of sugars in the endosperm of developing seeds of oilseed rape. Plant Physiology. 2003;**131**(1):228-236

Chapter 4

The Magical Chemistry and Nutritional Importance of Rice Seeds: A Review

Suman Sangwan, Harshita Singh, Susheel Gulati, Lalita Singh, Archana Malik and Suryapal Singh

Abstract

The most significant food on earth is rice. The nutrients included in rice include thiamine, riboflavin, niacin and tocopherol, as well as protein, fat, crude fiber, carbohydrates and minerals. It contributes significantly to human health by preventing diseases like high blood pressure, cancer, Alzheimer's, heart disease, skin conditions and dysentery. As a result, rice is an excellent option for natural sources of antioxidants and other therapeutic characteristics, and it may have the capacity. The magical chemistry enlightens the insights of functional groups, which makes it a potent food. The review also discusses how the nutritional content of rice seed changes as it ages.

Keywords: rice seed, nutritional importance, chemistry and human health, bioactive compounds, antioxidants

1. Introduction

Over 50% of people on earth eat rice as their primary meal. In 17 Asian and Pacific island nations, 9 North and South American nations, and 8 African nations, it is the main dietary energy source. In comparison to wheat's 19% and maize's (corn's) 5% contributions to the world's dietary energy supply, rice contributes 20% [1]. The amount of total free phenolics and total free flavonoids in the growing rice grains is rather high, which is connected well with their reduced capacity. The growing rice grains would be a rich source of phytochemicals because they have significant levels of free and soluble-ester ferulic acids. For billions of people worldwide, rice is a staple diet. It is a valuable source of fiber, carbs and numerous other nutrients. Humans have been eating rice since the Stone Age. Chinese archeologists discovered the earliest rice in 10,000 BC [2]. A cereal grass with nodes and internodes, rice develops as upright stems. It has ovate-acuminate, parallel-vented leaves that are oriented lengthwise and have short petioles. Two bracts that are connected at their roots form a "keel" around the spikelets, which are surrounded by two bracts [2]. Endosperm, bran and germ are the three components of a rice grain [2]. The grain itself is high in starch but low in fat or protein, but it still has all of the essential amino acids that humans require for nourishment; *The World Factbook 2018* [3].

Rice seeds are tiny, oval-shaped grains that sprout on many long-grass species of the Poaceae (or Gramineae) family of grasses, including *Oryza sativa*, *Oryza nivara*, *Oryza rufipogon*, etc. Today, around 125 types of rice are cultivated throughout the world, each with unique properties and applications depending on their botanical categorization. Jasmine rice, which is sourced from Asia, is the most popular type of rice consumed worldwide. Basmati rice is sourced from India, and arborio rice is sourced from Italy. Indica, vannarica, paragrass (*Vaccinium*), and other varieties of rice are also produced all over the world. About 40% of cooked rice's weight is made up of rice bran, which is highly nutritious. Starch, protein, fat and dietary fiber are the main ingredients of rice bran [4]. Rice bran contains a wide range of antioxidants including phenolic compounds such as flavonoids and tannins which have antioxidant properties [5]. Due to their ability to neutralize free radicals that harm cellular structures, phenolic compounds are regarded as potent antioxidants [6]. In addition to phenolics, rice bran also contains other types of antioxidants such as carotenoids which are responsible for providing yellow coloration in cooked rice products [7].

Rice seed is a very important part of the rice plant. It contains many essential nutrients for human consumption, and it also has many health benefits. Rice seed is composed of two major components: the endosperm and the bran. The bran consists mainly of pectin, cellulose, hemicellulose, lignin and some proteins, while the endosperm consists mainly of starch but also contains other compounds such as oil globules, amylopectin and others.

Rice seeds contain three different types of chemical substances: phytochemicals (substances generated by plants), secondary metabolites (substances created by microbes) and xenobiotics (compounds not produced by plants or microorganisms). Phytochemicals are seen as being more significant than secondary metabolites or xenobiotics since their positive effects on health outweigh their negative ones. Secondary metabolites can cause toxic effects if consumed in high amounts, but this is not true for phytochemicals since they have beneficial effects even when consumed in high amounts. Carotenoids have also been identified in rice seed (Oryza *sativa*). The primary carotenoid in rice is beta-carotene, which gives rice grains their orange hue. Beta-cryptoxanthin, gamma-carotene and alpha-carotene are other carotenoids that are present [8]. Beta-cryptoxanthin, which is abundant in carrots and is known for its capacity to absorb light energy, has been found to be a potent antioxidant [9]. Rice is a very nourishing food that is full of many vital nutrients like fiber, carbs, vitamins and minerals. More than half of the daily requirements of the most important nutrients for humans are provided by rice, which is also a fantastic source of energy [10]. It has been demonstrated that they possess antioxidant qualities, which may aid in shielding the body from harm brought on by free radicals. Highly reactive chemicals known as free radicals have the potential to harm biological tissues and cells. Antioxidants function as free radical scavengers, preventing cell damage and preserving the body's health.

2. Health benefits of Rice

Rice is a very healthy food that can provide many health benefits for those who consume it regularly. Some of the health advantages of eating rice include the following: lowering cholesterol fibre found in rice contributes to a reduction in blood cholesterol levels. Enhanced Digestion Increased consumption of fibre-rich foods, such as brown rice, will help digestion since they will keep you fuller longer, causing you to

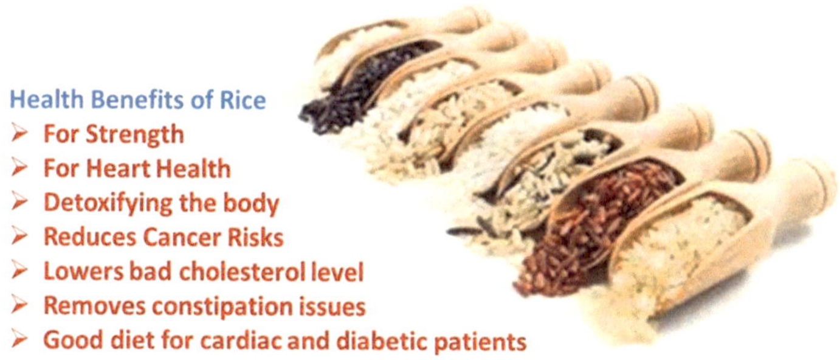

Figure 1.
Health benefits of different in rice seeds.

consume fewer calories than usual. This means that increasing your intake of whole grains will prevent your weight from rising as quickly because you will consume fewer calories than normal, which also lowers your risk of heart disease. The health benefits of rice and the avoidance of diseases including high blood pressure, cancer, skin care issues and dysentery are significant. Even tiny amounts of red rice can help you lose weight. It keeps us full. **Figure 1** and **Table 1** show the health benefits and nutrient content of rice per 100 g dry weight, respectively.

3. Chemistry of rice seeds

Starch, protein and fiber are all present in the rice grain on their own. The starch level, which has roughly 70% of its calories from carbohydrates, is the primary factor that makes rice so nutritious. When combined with other nutrients like protein and fiber as part of a balanced diet, this carbohydrate content gives the body energy. Additionally, rice contains significant amounts of manganese, which supports bone health by preserving strong bones and teeth. According to research, rice contains phenolic, polyphenolic, flavonoid, anthocyanin, anthocyanidin, tannin, vitamin E, tocopherol, tocotrienol, oryzanol, ferulate, phytic acid, phytate and other compounds that are crucial to the bioactivity of rice seeds [13].

The phenolic chemicals are present in the grain's cell wall in both a free form and an insoluble form. Phenolic chemicals can be divided into mono- and polyphenolics based on their structural similarities. It has been demonstrated that phenols contain antioxidant capabilities that assist shield cells from damage brought on by free radicals. Highly reactive chemicals known as free radicals have the potential to harm cells if they are not eliminated from the body. Antioxidants stop these dangerous chemicals from harming your body by removing them before they can cause any harm. Phenols have been demonstrated to lessen lipid peroxidation, which is what causes rancidity and is excellent for food preservation. The phenolic compounds gallic acid, protocatechuic acid (PCA) and vanillic acid are abundant in rice seed [13].

Rice seed growth is significantly influenced by flavonoids. There are a lot of phenolic substances in rice seeds, including tannins, phenolic acids, flavonoids and stilbenes [13]. Both antioxidant and anti-inflammatory properties are present in flavonoids. Because quercetin is one of the most prevalent phenolic chemicals in rice seed, we will concentrate on it in this study. It has been suggested that quercetin plays a role in the germination, maturation and production of endosperm in the

Nutrients	White rice	Brown rice	Red rice
Water Content %	12.7	37.6	12.4
Energy (KJ)	1736	1548	1426
Protein (g)	8.1	10.4	10.49
Fat (g)	0.8	0.9	0.81
Carbohydrate (g)	91	77.24	70.19
Fiber (g)	1.5	3.2	2.71
Sugar (g)	0.1	0.85	1.25
Calcium (mg)	32	23	18.71
Iron (mg)	0.91	1.47	13.45
Magnesium (mg)	28	143	192.27
Phosphorus (mg)	131	333	297
Potassium (mg)	131	223	128
Sodium (mg)	6	7	4
Zinc (mg)	1.24	2.02	1.91
Copper (mg)	0.25	1.2	0.8
Manganese (mg)	1.24	3.74	1.77
Selenium (μg)	17.2	23.4	18.19
Vitamin C (mg)	0.0	0.0	0.0
Thiamine (mg)	0.08	0.40	0.21
Riboflavin (mg)	0.06	0.09	0.05
Niacin (mg)	1.82	5.09	4.22
Pantothenic acid (mg)	1.15	1.49	1.72
Vitamin A (IU)	0	0.3	0.5
Vitamin E (mg)	0.13	0.9	1.2
Beta-carotene (μg)	0.1	1.2	1.5
Folate (μg)	9	20	23
Saturated fatty acids (g)	0.20	0.2	0.21
Polyunsaturated fatty acids (g)	0.24	0.21	0.31

Here: KJ, Kilo Joule; g, gram; mg, milligram; μg, microgram; IU-International Unit.

Table 1.
Nutrient content of rice per 100 g dry weight [11, 12].

development of rice seeds [14]. The relative orientation of different moieties in a molecule's chemical structure determines the metabolic actions of flavonoids and their metabolites. Following absorption, flavonoids are converted to smaller phenolic compounds in the liver by a process of glucuronidation, sulfation and methylation [14]. **Figures 2** and **3** enlighten the chemical structure of nutritional components and their health importance, respectively. The relative orientation of different moieties in a molecule's chemical structure determines the metabolic actions of flavonoids and their metabolites. Following absorption, flavonoids are converted to smaller phenolic

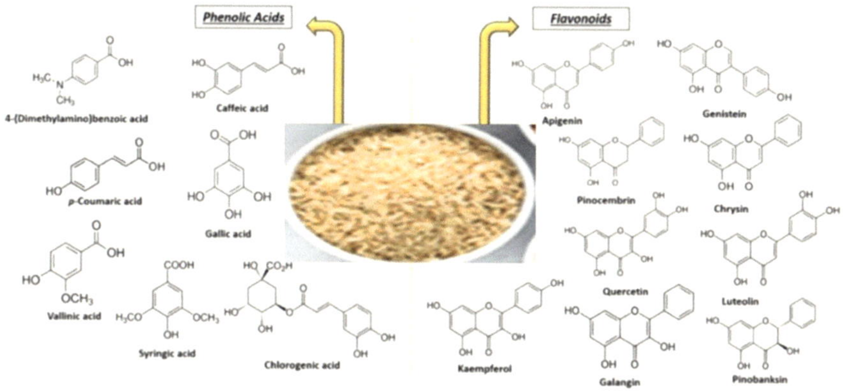

Figure 2.
Chemical structures of phenolic and flavonoids present different in rice seeds.

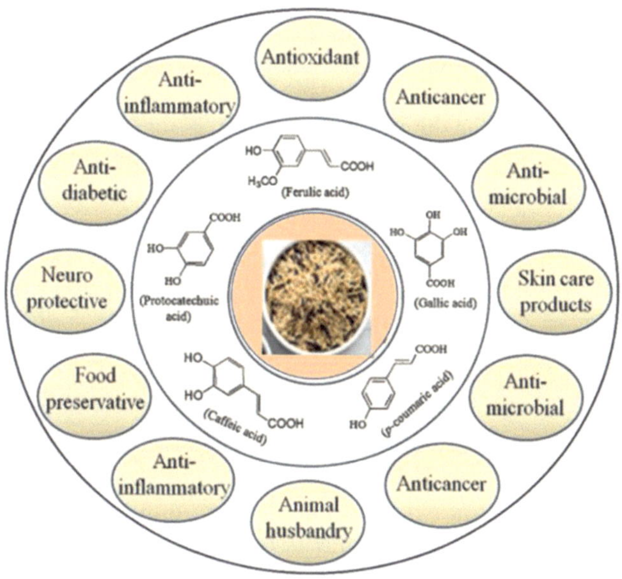

Figure 3.
Health importance of nutrients present in different rice seeds.

compounds in the liver by a process of glucuronidation, sulfation and methylation [15]. In addition, free radicals including super-oxide anion (NO) and peroxynitrite play important roles in the inflammatory process [16].

4. Changes in chemical properties of rice seeds during aging

Rice qualities like hydration, swelling, solubility, viscosity and pasting alter as it ages. The aging process has an impact on the physical and chemical characteristics of rice, primarily the physicochemical characteristics that affect cooking and pasting. Due to changes in the components and interactions between the components like protein, lipid and starch, aging also has an impact on the quality and usefulness of rice. Although there is no change in the protein content during storage, the solubility of rice in water decreases as a result of the intrinsic albumin's decreased solubility [17]. Contrarily, free amino acids in rice rise with storage [18]. Thus, the higher molecular weight peptides are increasing while the lower molecular weight peptides are decreasing [19].

The research stated that even after 13 weeks of storage, rice's amylose and amylopectin quantities do not appreciably change [20]. Some literature has been found that an increased water-insolubility of starch and protein in rice with aging, resulting in a slower rate of cooking as α-amylase and β-amylase in rough rice decrease significantly during storage [17]. Due to the hydrolysis of lipids to form free fatty acids and the oxidation of lipids that may result in hydroperoxides, changes in the lipid profile of rice occurred during storage [17]. Free phenolic acids from the rice grain are released during storage, resulting in the creation of free fatty acids (FFAs), which deplete important antioxidants. It has been discovered that hydroperoxides, carbonyl compounds and amylase all interact with FFAs. These processes hasten the oxidation and condensation of proteins, which results in the accumulation of volatile carbonyl chemicals in the grain [17]. As a result of aging, the texture of rice becomes harder and less sticky compared with fresh rice as it is a complicated process and involves physical, chemical and biological changes in rice. It is difficult to address the combined consequences, and it is difficult to determine the paddy grains' nutritional value and food safety. The hydrolysis and oxidation of lipids to free fatty acids or peroxides, which cause the increase in acidity and have a substantial impact on the flavor of rice, are also linked to the corrosion of rice flavor [17]. Research has also shown that rice seeds stored at ambient temperature show a significant change in textural properties over those stored at cold temperature [21], and remarkably, Tananuwong and Malila; 2011 [22] observed that after 12 months of storage at ambient conditions, hardness of hulled rice, when cooked, increases.

5. Conclusion

Rice is a powerhouse of antioxidants, vitamins and minerals like zinc, calcium, magnesium, zinc and sodium, which contribute greatly to overall well-being. Rice is also low on glycaemic index, which means that it can control blood sugar. The coronavirus pandemic has made us realize the importance of lung health like never before. With the contagious infection still lurking around, people with a history of lung diseases like asthma and respiratory problems are at high risk. Including rice in the daily diet improves lung capacity, with its copious amounts of magnesium and selenium. Rice not only improves and regulates breathing patterns but also improves oxygen consumption and circulation to each cell in the body. The aging impacts on rice seed quality enlighten future research and may further the investigation of techniques and alternatives to achieve high-quality rice seeds with minimal processing and for preservation of essential quality traits.

Author details

Suman Sangwan[1], Harshita Singh[2], Susheel Gulati[1], Lalita Singh[3], Archana Malik[4] and Suryapal Singh[5*]

1 Department of Chemistry, Chaudhary Charan Singh Haryana Agricultural University, Hisar, India

2 Department of Vegetable Sciences, Chaudhary Charan Singh Haryana Agricultural University, Hisar, India

3 Department of Botany, Maharshi Dayanand University, Rohtak, India

4 Department of Chemistry, Dayanand College, Hisar, India

5 Department of Seed Sciences, Chaudhary Charan Singh Haryana Agricultural University, Hisar, India

*Address all correspondence to: surajahlawat06@gmail.com

References

[1] Esa N, Rajamani L, Yusoff ZM. Reengineering Local Knowledge: Life. Science and Technology (Penerbit USM): Penerbit Usm; 2014

[2] Carriger S, Vallée D. More crop per drop. Rice Today. 2007;**6**(2):10-13

[3] Nguyen NH. In: Nguyen NH, editor. La evolución de the World Factbook 2018 en español: The Evolution of the World Factbook 2018 In Spanish. 2018

[4] Burman M, Nair SK, Sarawgi AK. Principal component analysis for yield and its attributing traits in aromatic landraces of rice (*Oryza sativa* L.). International Journal of Bio-resource and Stress Management. 2021;**12**(4):303-308

[5] Kumar N, Goel N. Phenolic acids: Natural versatile molecules with promising therapeutic applications. Biotechnology Reports. 2019;**24**:e00370

[6] Qin F, Yao L, Lu C, Li C, Zhou Y, Su C, et al. Phenolic composition, antioxidant and antibacterial properties, and in vitro anti-HepG2 cell activities of wild apricot (Armeniaca Sibirica L. lam) kernel skins. Food and Chemical Toxicology. 2019;**129**:354-364

[7] Andriani R, Subroto T, Ishmayana S, Kurnia D. Enhancement methods of antioxidant capacity in rice bran: A review. Food. 2022;**11**(19):2994

[8] Miller AP, Coronel J, Amengual J. The role of β-carotene and vitamin a in atherogenesis: Evidences from preclinical and clinical studies. Biochimica et Biophysica Acta (BBA)-molecular and cell biology of. Lipids. 2020;**1865**(11):158635

[9] Coronel J, Pinos I, Amengual J. β-Carotene in obesity research: Technical considerations and current status of the field. Nutrients. 2019;**11**(4):842

[10] Rao ND, Min J, DeFries R, Ghosh-Jerath S, Valin H, Fanzo J. Healthy, affordable and climate-friendly diets in India. Global Environmental Change. 2018;**49**:154-165

[11] Muttagi GC, Ravindra U. Chemical and nutritional composition of traditional rice varieties of Karnataka. Journal of Pharmacognosy and Phytochemistry. 2020;**9**(5):2300-2309

[12] Kowsalya P, Sharanyakanth PS, Mahendran R. Traditional rice varieties: A comprehensive review on its nutritional, medicinal, therapeutic and health benefit potential. Journal of Food Composition and Analysis. 2022:104742

[13] Corso M, Perreau F, Mouille G, Lepiniec L. Specialized phenolic compounds in seeds: Structures, functions, and regulations. Plant Science. 2020;**296**:110471

[14] Wisetkomolmat J, Arjin C, Hongsibsong S, Ruksiriwanich W, Niwat C, Tiyayon P, et al. Antioxidant activities and characterization of polyphenols from selected northern Thai Rice husks: Relation with seed attributes. Rice Science. 2023;**30**(2):148-159

[15] Xiao J, Chen T, Cao H. Flavonoid glycosylation and biological benefits. Biotechnology Advances: S0734-9750 (14)00092-5. 2014. DOI: 10.1016/j.biotechadv.2014.05.004

[16] Wongsa P. Phenolic compounds and potential health benefits of pigmented rice. Recent Advances in Rice Research. 2020;**4**:19-21

[17] Zhou Z, Robards K, Helliwell S, Blanchard C. Ageing of stored rice:

Changes in chemical and physical attributes. Journal of Cereal Science. 2002;**35**(1):65-78

[18] Dhaliwal YS, Sekhon KS, Nagi HPS. Enzymatic activities and rheological properties of stored rice. Cereal Chemistry. 1991;**68**(1):18-21

[19] Chrastil J. Chemical and physicochemical changes of rice during storage at different temperatures. Journal of Cereal Science. 1990;**11**(1):71-85

[20] Abeysundara AT, Navaratne SB, Wickramasinghe I, Ekanayake D. Determination of changes occurrence in important physical properties of paddy during early storage. 2017

[21] Wiset L, Laoprasert P, Poomsa-ad N, Tulyathan V. Effects of in-bin aeration storage on physicochemical properties and quality of glutinous rice cultivar RD 6. Australian Journal of Crop Science. 2011;**5**(6):635-640

[22] Tananuwong K, Malila Y. Changes in physicochemical properties of organic hulled rice during storage under different conditions. Food Chemistry. 2011;**125**(1):179-185

Chapter 5

Seed Production and Handling of Two Important Conifers Grown in Kenya

Peter Murithi Angaine, Alice Adongo Onyango and Jesse Owino

Abstract

Pinus patula and *Cupressus lusitanica* are key commercial forestry plantation species introduced in Kenya. There are many uses for these species in industry creating a huge demand for their products. The demand has caused increased plantation establishment with seed as the major source of propagule. The many developments in the forestry sector have led to the need for low energy rapid extraction technique that improve seed quantity and quality from the available sources. There have been developments in improving extraction and quality which offer an opportunity for better seed collection and handling techniques for these conifers. This chapter will focus on improvement of seed production and handling of the two conifers that aids in the design of low energy-intensive methods that reduce the duration for extraction, optimize seed yield and enhance seed quality.

Keywords: *Pinus patula*, *Cupressus lusitanica*, cone and seed characteristics, seed extraction, seed yield, germination

1. Introduction

Pinus patula Schiede & Deppe belongs to the Pinaceae family, but *Cupressus lusitanica* Mill. belongs to the Cupressaceae family. Coniferous gymnosperms make up both species [1–3]. Mexican cypress, also known as *C. lusitanica*, was brought to Kenya in 1936, but it was not until the 1950s that it was widely planted to replace *Cupressus macrocarpa*, which had been introduced earlier but had developed a sensitivity to Cypress Canker [4, 5]. The two primary commercial forest plantation species utilized to supply saw-wood to Kenyan forest industries for the manufacturing of wood for furniture and construction, round wood for plywood, and fiber to produce pulp and paper are *C. lusitanica* and *P. patula* [6–8].

Kenya Forest Service (KFS) is a corporate body that provides for the development and sustainable management, including conservation and rational utilization of all forest resources for the socioeconomic development of Kenya [9]. Gazetted forest land consists of about two thirds of protected forest reserve that comprise indigenous tree species, and about 150,000 ha of exotic softwood plantations [5, 10] (**Figure 1**).

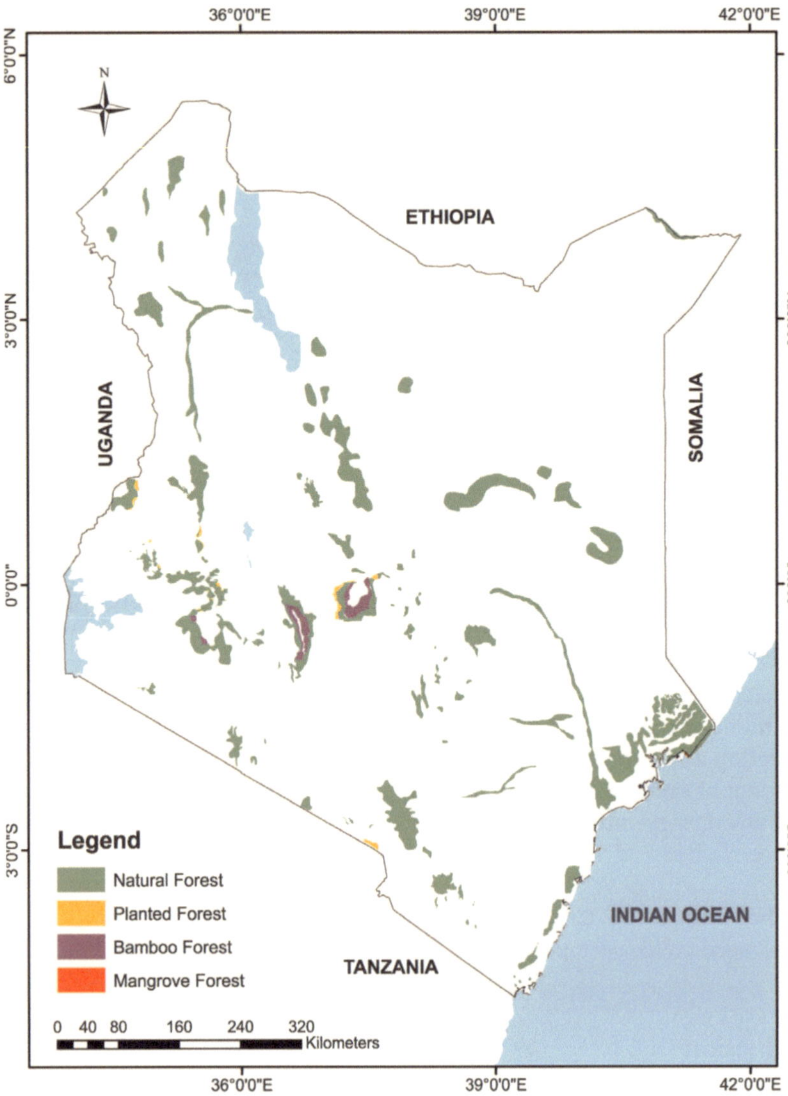

Figure 1.
Pine and cypress plantations fall under "Planted Forest" area [11].

C. lusitanica is the most widely planted species occupying about 55% of the plantation area, and *P. patula* taking 25% [12]. The agency that provides seeds for forest tree species is Kenya Forestry Research Institute (KEFRI) [13].

Despite sourcing seeds of *P. patula* and *C. lusitanica* from selected seed stands and seed orchards [14, 15], the demand for seed is not being met. Currently, the demand for *P. patula* seed in Kenya stands at 1000 kg, of which KEFRI can only supply 600 kg, while *C. lusitanica* seed is 1500 kg [16]. The bulk of this seed is taken by KFS for their annual planting programme, with the remainder going to private commercial nurseries and farmers. Taking into account the effects of climate change and the limited number of seed sources, it is essential to review seed collection and processing practices, to ensure improved quality and quantity of seed [17, 18]. Consequently, there are numerous opportunities to be explored in this regard. Therefore, opportunities

DOI: http://dx.doi.org/10.5772/intechopen.1002322

exist on quality improvement in terms of where to collect cones within the crown, the size of cones to be collected and how to improve seed yield through seed processing practices that will impact germination rate and packaging.

The gaps in the improvement of *P. patula* and *C. lusitanica* seed yield have been identified as the need to understand the influences of: (i) crown morphometry on seed production, (ii) cone characteristics on seed production, and (iii) extraction practices on seed yield and germination. To address these gaps, this chapter focuses on recent developments in collection and handling of *P. patula* and *C. lusitanica* seeds.

2. Materials and methods

2.1 Study site

The study materials were sourced from seed orchards in Londiani, Kenya. Londiani area is at an elevation of 2308 m.a.s.l. with an average temperature of 15.7°C (minimum of 8.6°C and maximum of 23.31°C). The area has two rainy seasons occurring in the months of March to May with an average rainfall of 750 mm and in October to December with average rainfall of 423 mm. The driest months are January to February and August to September [7, 19–21].

2.2 Sampling frame

2.2.1 Pinus patula

Crown morphometry and seed production: the sampling procedure was adopted from the method used by [21, 22]. The orchard was broken into three equal segments and 10 trees were picked and measured for their diameter at 130 cm (D130) and height from each area. One tree, which had produced the most fruit and had the greatest diameter (D130), was then selected and marked for cone collection from each of the three plots. Measurements were taken for D130 (cm), tree height (m), crown height (m) and crown radius (m) = D(m)/2 for the chosen trees (**Figure 2**). The crown of the tree was then divided into three equal parts; top (A), middle (B) and bottom (C). It was further split into two sections based on the distance from the stem (0–2 m = 1, >2 m = 2). Section 1 covered the part that was 2 m away from the stem (A1, B1, and C1), while Section 2 was made up of the part greater than 2 m from the stem (A2, B2, and C2) [15, 21]. From each crown section, 15 ripe cones were taken, amounting to 90 cones per tree. The gathered cones were put in individual bags for every section of the crown and then transported to the laboratory.

P. patula cones were assigned a particular identity based on the tree and crown sector they were taken from, with a maximum of 10 cones per sector. Before they were extracted, the characteristics of each cone were observed, including shape (straight or curved), length (in cm), diameter at the widest point (in cm), and weight (in g). The cones were placed on glass petri dishes and heated in an oven at 65°C for 24 hours. The seeds were extracted by gently tapping the cones 15 times on a flat wooden bench. The weight of the cones without the seeds and the total number of seeds per cone were recorded. The percentage of cones that opened after heat treatment for seed extraction was also measured.

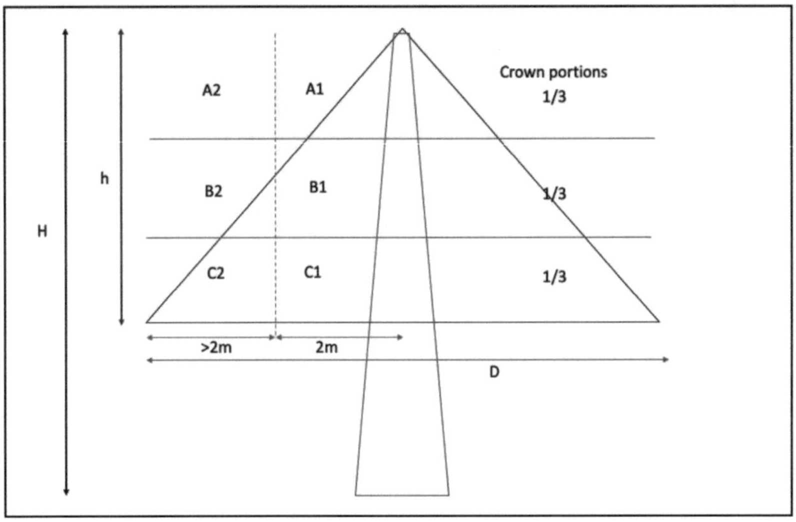

Figure 2.
Schematic diagram on crown compartmentalization for sampling area of Pinus patula *cones.*

2.2.2 Cupressus lusitanica

The study materials were taken from a 14-year-old clonal seed orchard of *Cupressus lusitanica* in the Londiani region. Thirty trees that were producing seeds were randomly chosen and fifty mature cones were taken from each, giving a total of 1500 cones. Cones that were brown and shut at the time of collection were judged to be mature [23]. The cones were bundled together and put in gunny bags before being brought to the laboratory for seed extraction, screening and weighing.

2.3 Experimental design

2.3.1 Pinus patula

Cone characterization was based on the method described by [21]. This involves categorizing mature cones according to their shape (straight or curved) (**Figure 3**), length (L1) (cm), diameter at widest part (cm) and weight (g). The cones were then placed on uncovered glass Petri dishes and heated in an oven (Yamato DS411) at 65°C for 24 hours [19]. After drying, the cones were removed and seeds extracted by tapping 15 times on a flat wooden bench. Measurements of the length of the part of the cone that opened (L2) (cm) (**Figure 4**), the weight of the cone without seed (g) and the total seed count from each cone were taken. The percentage of the cone that opens after drying for seed extraction (p) was calculated using Eq. (1); this p is used to compare the opening length of the cone in relation to its shape [19, 21, 24].

$$p = \frac{L2(cm)}{L1(cm)} * 100 \qquad (1)$$

Cone pretreatment and seed extraction: the cones were identified and measured for length, diameter and weight, then put in hot (100°C) and room temperature

Figure 3.
Pinus patula *cones ranked based on shape (straight-top and curved-bottom, with a 30 cm ruler on the side showing scale).*

Figure 4.
Pinus patula *cones showing stages of cone opening.*

(25°C) water for 10 minutes, 24 hours and a control (not soaked) to examine the impact of humidity [19, 25]. The cones were arranged in labeled glass Petri dishes with sufficient distance between them to avoid cross-contamination of the seeds, and then exposed to artificial heating for seed extraction. Eight temperatures (30, 40, 50, 65, 70, 75 and 85°C) [26] and a DB condition (drying bed to simulate real-life seed extraction with a temperature of 44.8 ± 6.00°C) were tested for 4, 24 and 48 hours together with the control (no soaking).

2.3.2 Cupressus lusitanica

Cone and seed characterization: the cones were divided into two categories, large (CB) and small (CS), based on sieving with a 20 mm sieve. After assessing for maturity and any defects, 240 defect-free mature cones were chosen from each category. The diameter and weight of each cone was measured. Then, the cones were heated artificially at 65°C for 48 hours to extract the seeds [19]. The seeds were then sorted by sieving using a 2 mm sieve to create small (SS) and large (BS) category. To further categorize the seeds, they were floated in water for 5 minutes, and divided into floaters (FF) and sinkers (SS) (**Figure 5**). Finally, the seeds were germinated according to their categorization.

Cone pretreatment and seed extraction: In order to conduct an experiment, 540 cones were randomly divided into three groups of 180 cones each. The first group acted as the control and was not soaked in any liquid. The second and third groups were submerged in cold and hot water, respectively, for 10 minutes [19, 25, 27]. The three groups were further broken down into 6 different temperature categories (GH, DB, 40, 50, 65, and 85°C) of 30 cones each and labeled. The categories of 'greenhouse' (GH) and 'seed drying bed' (DB) refer to two infrastructures used for seed extraction in Londiani, Kenya.

Observations for seed release and seed counts per cone were done at 24 hours and at 48 hours. Seeds that had not been released from open cones were forcefully removed manually using forceps to account for seeds that were retained in a cone and labeled as 48F (48 hours forced).

Germination tests were conducted based on the extraction treatments. Radicle emergence was taken as the criterion for germinability [28].

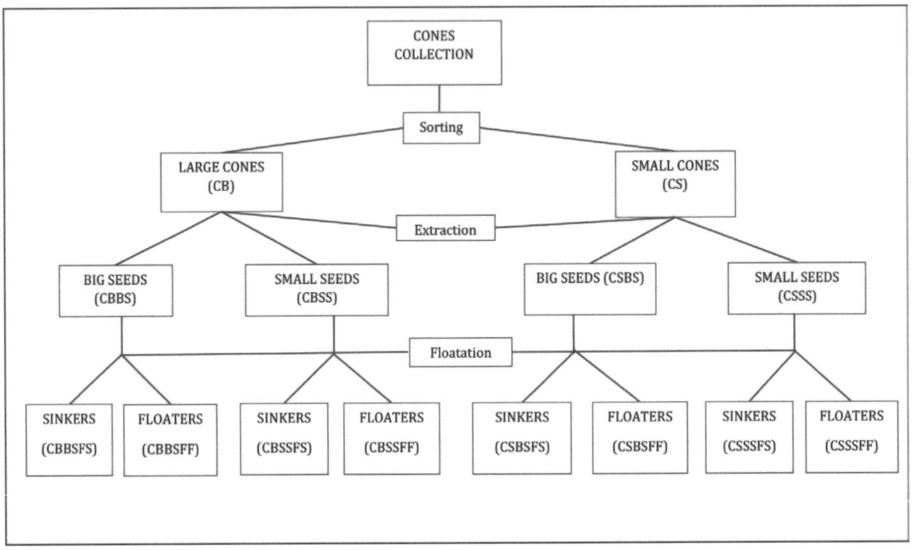

Figure 5.
Experimental layout for Cupressus lusitanica *cones and seed categorization [20].*

3. Results and discussion

3.1 *Pinus patula*

The different sections of the crown had varied seed yield, with compartment A2 missing from the crown due to the conical shape of *P. patula* crown. Compartment A1 had the highest number of seeds (33.3 ± 4.91) from the cones, while C2 had the lowest (14.4 ± 2.76). This is in agreement with research done on another species, *Pinus densiflora* [29]. Compartment C2 had the highest percentage of opening (46.6 ± 1.98%), yet this did not result in a high seed yield (**Figure 6**). This study found that neither cone shape nor percent opening had an effect on seed yield [21]. Mathematical modeling was used to analyze the results, which showed that the length of the cone had the greatest influence on seed yield [24].

Research has indicated that cone soaking does not have a major influence on cone opening. Temperature of extraction was seen to have a major effect on cone opening and the subsequent seed yield, with higher temperatures favoring both cone opening

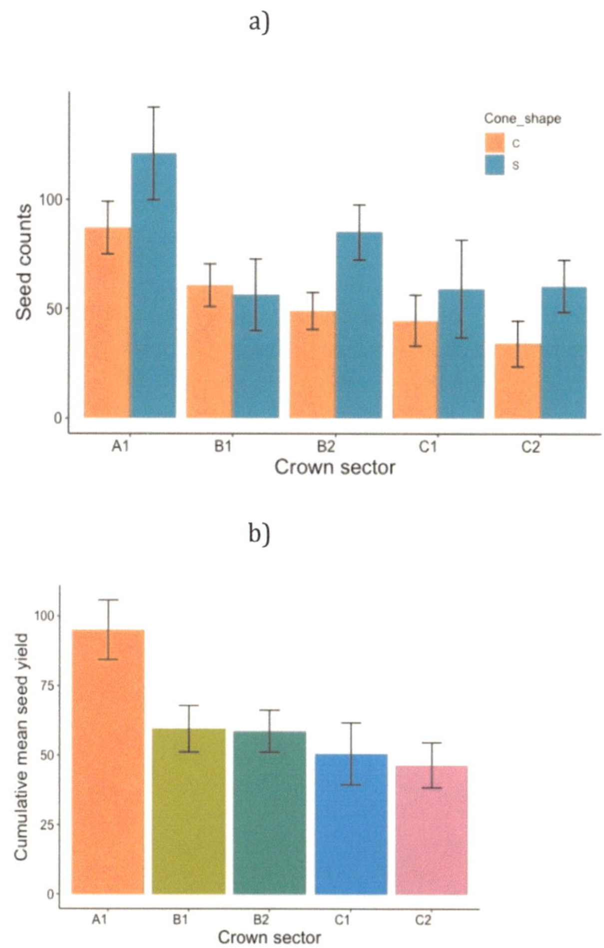

Figure 6.
a) Seed yield by cone shape and crown compartment and b) Cumulative mean seed yield by compartment.

and seed yield [19]. Soaking of cones increases the moisture content of the cone, which can hinder the swift opening of the cones [19, 30, 31].

3.2 *Cupressus lusitanica*

Sieving of cones and seeds was found to be a helpful tool when it comes to predicting seed yield and quality. It was observed that bigger seeds from both small and big cones had better germination performance. Floatation was found to be an effective method of predicting germination, where denser (sinkers) seeds yielded better results than floaters. Big seeds yielded 95,000–105,000 seeds per kilogram, with a germination rate of 51%. This is a big improvement from the previous rate of 25% and 160,000–290,000 seeds per kilogram [20, 32, 33]. Different from pines, soaking cones in cold water prior to seed extraction was found to have a significant effect on cone opening and seed release, with 77% of seeds released within the first 24 hours, 90% within 48 hours and 10% having to be forced out [2, 34]. Temperature was also seen to affect seed release, with higher temperatures leading to higher seed yield than lower temperatures [35, 36]. After 30 days of assessment, it was observed that germination performance was highest for seeds extracted from cones that had been soaked in cold water. Several studies have been conducted on the effects of soaking media on seed germination, with the present study focusing on soaking cypress cones in cold and hot water and their influences on germination [37–40].

4. Conclusions

The upper sections of *P. patula,* crown should be targeted during seed collection for higher seed yield. Larger cones for both *P. patula and C. lusitanica* have highest seed yield. Combined effect of soaking and higher temperature exposure of *P. patula* and *C. lusitanica* cones have a positive effect on seed yield without adverse effects on seed germination.

Acknowledgements

The authors acknowledge the Kenya Forestry Research Institute, for facilitating data collection and experimental phase for this study.

Conflict of interest

The authors declare no conflict of interest.

Author details

Peter Murithi Angaine*, Alice Adongo Onyango and Jesse Owino
Kenya Forestry Research Institute, Rift Valley Ecoregion Research Centre, Londiani, Kenya

*Address all correspondence to: pangaine@gmail.com

References

[1] Matziris D. Genetic variation in cone and seed characteristics in a clonal seed orchard of Aleppo pine grown in Greece. Silvae Genetica. 1998;**47**(1):37-41

[2] Barbour MG. Closed-cone pine and cypress forests. In: Barbour MG, Keeler-Wolf T, Schoenherr AA, editors. Terrestial Vegetation of California. 3rd ed. Los Angeles: University of California Press; 2007. pp. 296-312

[3] Fernando DD. The pine reproductive process in temperate and tropical regions. New Forest. 2014;**45**(3):333-352

[4] Mbinga J, Omondi SF, Onyango AA. Conifers: Species diversity and improvement status in Kenya. In: Conifers-Recent Advances. London, UK: IntechOpen; 2021

[5] Kuria NC, Balozi KB, Kipkore W. Growth and yield models for plantation-grown *Cupressus lusitanica* for Central Kenya. African Journal of Education, Science and Technology. 2019;5(2):34-58

[6] Kindt R, Dawson IK, Muchugi A, Pedercini F, Roshetko JM, Van Noordwijk M, et al. The one hundred tree species prioritized for planting in the tropics and subtropics as indicated by database mining. In: Working Paper No. 312. Vol. 312. Nairobi, Kenya: World Agroforestry; 2021. pp. 7-13

[7] Ngugi MR, Mason EG, Whyte AGD. New growth models for Cupressus *lusitanica* and *Pinus patula* in Kenya. Journal of Tropical Forest Science. 2000; **12**(3):524-541

[8] *Pinus patula* (PROTA). (2016, January 19). PlantUse English. Retrieved 06:06, September 7, 2023. Available from: https://uses.plantnet-project.org/e/index.php?title=Pinus_patula_(PROTA)&oldid=201393

[9] Go K. Forest Conservation and Management Act. Kenya Gazette Supplement. 2016;**155**(34):677-736

[10] Mathu WJK. Growth, Yield and Silvicultural Management of Exotic Timber Species in Kenya [Internet]. University of British Columbia; 1983. Available from: https://open.library.ubc.ca/soa/cIRcle/collections/ubctheses/831/items/1.0075426

[11] Ototo G, Vlosky RP. Overview of the forest sector in Kenya. Forest Products Journal. 2018;**68**(1):6-14

[12] Ho KV, Kröel-Dulay G, Tölgyesi C, Bátori Z, Tanács E, Kertész M, et al. Non-native tree plantations are weak substitutes for near-natural forests regarding plant diversity and ecological value. Forest Ecology and Management. 2023;**531**:120789

[13] Cheboiwo JK. Public private partnerships opportunities for forestry sector development in Kenya: Synthesis of primary and secondary production actors, and trade. Journal of Environment and Earth Science. 2018;**8**(1):47-69

[14] Pérez-Luna A, Wehenkel C, Prieto-Ruíz JÁ, López-Upton J, Solís-González S, Chávez-Simental JA, et al. Grafting in conifers: A review. Pakistan Journal of Botany. 2020;**52**(4):1369-1378

[15] Bilir N, Prescher F, Lindgren D, Kroon J. Variation in cone and seed characters in clonal seed orchards of *Pinus sylvestris*. New Forest. 2008;**36**(2):187-199

[16] Njoroge G. Towards a Framework for the Utilisation of Forest Resources: A Study of Nakuru District. Mendeley: University of Nairobi; 1986

[17] Valette M, Vinceti B, Gregorio N, Bailey A, Thomas E, Jalonen R. Beyond fixes that fail: Identifying sustainable improvements to tree seed supply and farmer participation in forest and landscape restoration. Ecology and Society. 2020;**25**(4):1-26

[18] Avana-Tientcheu M, Marunda C, Fongnzossie E, Kemeuze V, Ngeuguim J, Mutta D. Tree germplasm management systems and their potential for sustaining plantation forestry in West and Central Africa. African Journal of Rural Development. 2019; **4**(1):33-63

[19] Onyango AA, Angaine PM, Inoti SK, Owino JO. Patula pine (Pinus patula) cones opening under different treatments for rapid seed extraction in Londiani. Kenya Journal of Horticulture and Forestry. 2020;**12**(2):63-69

[20] Angaine PM, Onyango AA, Ndungu SM, Inoti SK, Owino J. Influence of Cupressus lusitanica Mill. cones and seed characterization on germination in Kenya. Journal of Forests. 2021;**8**(2): 123-130. Available from: http://www.pakinsight.com/archive/101/12-2021/2

[21] Angaine PM, Onyango AA, Owino JO. Morphometrics of *Pinus patula* crown and its effect on cone characteristics and seed yield in Kenya. Journal of Horticulture and Forestry. 2020;**12**(3):94-100. Available from: http://www.academicjournals.org/JHF

[22] Hess AF, da Silveira AC, Krefta SM, dos Santos DV, Filho MDHV, Atanazio KA, et al. Crown dynamics of Brazilian pine ('*Araucaria angustifolia*') in Santa Catarina region of Brazil. Australian Journal of Crop Science. 2018; **12**(3):449-457

[23] Raddi P, Danti R, Della Rocca G. x *Cupressocyparis leylandii*. In: Enzyklopädie der Holzgewächse: Handbuch und Atlas der Dendrologie. Wiley-VCH Verlag GmbH & Co. KGaA; 2019. pp. 1-17. DOI: 10.1002/9783527678518.ehg2019001

[24] Owino JO, Angaine PM, Onyango AA, Ojunga SO, Otuoma J. Evaluating variation in seed quality attributes in *Pinus patula* clonal orchards using cone cluster analysis. Journal of Forestry. 2020;**7**(1):1-8

[25] Aniszewska M. Changes in humidity and temperature inside the pine cones (*Pinus sylvestris* L.) in two stages seed extraction. Leśne Prace Badawcze (Forest Research Papers). 2013;**74**(3): 205-214

[26] Schmidt LH. Guide to Handling of Tropical and Subtropical Forest Seed. Vol. 2000. Humlebaek, Denmark: Danida Forest Seed Centre; 2000. pp. 1-12

[27] Lovreglio R, Salvatore R, Giaquinto P, Leone V. Thermal treatments and germination response over time of seeds from serotinous and non serotinous cones of *Pinus halepensis* Mill. In: International Workshop MEDPINE 3: Conservation, Regeneration and Restoration of Mediterranean Pines and their Ecosystems. 2007. pp. 155-166

[28] Matthews S, Powell A. Towards automated single counts of radicle emergence to predict seed and seedling vigour. Seed Testing International (ISTA Bulletin). 2012;**142**:44-48

[29] Iwaizumi MG, Ubukata M, Yamada H. Within-crown cone production patterns dependent on cone productivities in *Pinus densiflora*: Effects of vertically differential, pollination-related, cone-growing conditions. Botany. 2008;**86**(6):576-586

[30] Wyse SV, Brown JE, Hulme PE. Seed release by a serotinous pine in the absence of fire: Implications for invasion into temperate regions. AoB Plants. 2019;**11**(6):1-8

[31] Ghildiyal S, Sharma C, Gairola S. Effect of temperature on cone bursting, seed extraction and germination in five provenances of *Pinus roxburghii* from Garhwal Himalaya in India. Southern Forests: A Journal of Forest Science. 2008;**70**(1):1-5

[32] Albrecht J. Tree Seed Handbook of Kenya. In: Omondi W, Maua JO, Gachathi FN, editors. 2nd ed. Kenya Forestry Research Institute. Nairobi, Kenya: GTZ Forestry Seed Centre Muguga; 1993. Available from: https://www.worldcat.org/title/tree-seed-handbook-of- kenya/oclc/38979878

[33] *Cupressus lusitanica* (PROTA). (2015, October 11). PlantUse English. Retrieved 06:17, September 7, 2023. Available from: https://uses.plantnet-project.org/e/index.php?title=Cupressus_lusitanica_(PROTA)&oldid=199082

[34] Garcillán PP. Seed release without fire in Callitropsis guadalupensis, a serotinous cypress of a Mediterranean-climate oceanic island. Journal of Arid Environments. 2010;**74**(4):512-515

[35] Milich KL. Cone Serotinity and Seed Viability of Fire-Prone California Cupressus Species. Humboldt State University; 2010

[36] Lev-yadun S. Living serotinous cones in Cupressus sempervirens. International Journal of Plant Sciences. 1995;**156**(1):50-54

[37] Sfairi Y, Lahcen O, Najib M, Feddy A, Abbad A. Dormancy-breaking and salinity/water stress effects on seed germination of Atlas cypress, an endemic and threatened coniferous species in Morocco. African Journal of Biotechnology. 2012;**11**(19):4385-4390

[38] Fierro-Cabo A, Plamann A. Enhancing the seed germination process of Montezuma cypress (*Taxodium mucronatum* Ten.). Journal of Forest Research. 2021;**26**(1):81-85

[39] Hossain MA, Islam KS, Rajasree N, Hossain MK, Alam MS. Pre-sowing treatments for improved germination and growth of two rare native species of Bangladesh. Journal of Forest Research. 2018;**29**(5):1277-1282

[40] Porter RH. Recent developments in seed technology. The Botanical Review. 1949;**XV**(5):283-344

Chapter 6

The Influence of Planting Time on the Seed Yield and Quality Millet/*Panicum miliaceum*. L

Khishigbuyan Turbat, Gungaanyam Galkhvv and Namjilsuren Jamiyan

Abstract

Millet has been cultivated in Mongolia since ancient times, and some historians believe that this cultivation may be the first crop of nomadic people. Academician P.M. Zhukovsky noted that multiline varieties of millet were found in the mountainous regions of Central Asia and Mongolia. Therefore, one of the regions of millet is Mongolia. There are records that Mongolians cultivated small rice in the 8th–12th centuries and called it Mongolian grain, cron, black rice, and millet, used it for food, and sold it to traders in China and other countries. The research study was conducted in the research area of the Research Institute of Plant and Agriculture in the area of Khongor Sum, Darkhan-Uul Province, Mongolia, in 2017–2020. The millet variety of Saratovskaya-853 was planted for seeds on May 20, 30, and June 10 at the rate of 3 million seed/ha and with 3 repetitions each. According to our research, the 20th of May is the most profitable time for planting, with a yield of 23.7 tons/ha. During this period, the number of weeds in the field was not much, the amount of protein contained in the seeds is 0.9–1.6% more than other versions, the seeds were mature, they are not affected by cold shocks and frosts, and they form a good seed casting protection against the cold. Spacing between 0.5 m and 15 cm between plants/total 9 stages. According to our research, the 20th of May is the most profitable time for sowing, with a yield of 23.7 cents/ha. During this period, the number of weeds in the field is low, the amount of protein in the seeds is 0.9–1.6% higher than other options, the seed yield is good, and the conditions are not affected by cold shock. The conditions for an increase in the yield of the version of May 20 have been established.

Keywords: cultivation, field, version, repetition, protein

1. Introduction

Millets (*Panicum miliaceum*. L) are a highly varied group of small-seeded grasses, widely grown around the world as cereal crops or grains for fodder and human food [1]. Most species generally referred to as millets belong to the tribe Paniceae around the world, millet has been grown for food and fodder. *Panicum miliaceum* is a tetraploid species with a base chromosome number of 18, within the genus Panicum [2].

The millet is an annual food and annual plant. It is one of the ancient crops cultivated in the world more than 7000 years ago [3].

It is suitable for growing in regions with low fertility, especially India, the Sahara desert, and West Africa, where the average rainfall is usually less than 500 mm, and where the soil is sandy and slightly acidic [4].

The cultivation of millet worldwide is decreasing year by year. According to the World Food and Agriculture Organization (FAO), as of 2019, a total of 718 million hectares were planted in the world and 863 million tons of crops were harvested. As of 2020, the world's total millet crop is 30.5 million tons, of which India (42%), Nigeria (20%), China (6%), Niger (12%), Mali (5%), and Ethiopia (3%) occupying [2].

The cultivation of millet/*Panicum miliaceum. L*/worldwide is decreasing year by year. For example, India is the leading country in the production of small rice, and in 1970, the total harvest in the country met 100% of domestic needs, while in the late 1970s, this consumption decreased to 50–75% [5].

However, since 2005, small rice has become less of a staple food, and it is mainly cultivated for animal feed and alcoholic beverages. The United Nations General Assembly at its 75th session in March 2021 declared 2023 the International Year of Millets [2]. Millets can grow on arid lands with minimal inputs and are resilient to changes in climate. As a result of the International Year of Small Rice, it will be possible to use the resistant qualities of the grain to increase its nutritional value, to introduce this grain to future generations for better production, better nutrition, and to produce more food and feed [2].

Mongolia is located in the central part of Asia between $41^{0}31^{I}$-$52^{0}09^{I}$ north latitude, the average height above sea level is 1580 m, the lowest point is 560 m, and the highest point is 4347 m. It has an extreme climate, arid and arid, and it has unique features indicating the characteristics of a temperate region. Due to its elevation above sea level, it has a cool climate and annual precipitation is 250–350 mm.

The first cold shock of spring occurs in the middle ten days of May, and the first cold shock of autumn occurs in the first 10 days of September. Seasonal and daily temperature fluctuations are large, reaching +40°C in July and −40°C in January. The average daily temperature during plant growth is 16.5°C, suitable temperature for plant growth is 2274.716.5°C. Crops with a short growing season and suitable for high altitude and cool regions are usually cultivated.

An average of 400,000–600,000 hectares of land will be cultivated annually, and about 70% of it will be cultivated only with grain plants, mainly wheat. Dark chestnut and chestnut soils with sandy and mechanical compositions are prevalent. The soil climate is suitable for growing wheat and other grains, potatoes, and vegetables.

2. The purpose of the study

In the conditions of the central cultivation region of Mongolia, the following objectives have been proposed in order to detect the possible period of harvesting of millet seeds

1. To study the effect of planting time on field germination

2. Choosing the right time to harvest millet seeds

3. To study the effect of planting time on the biochemical quality of millet seeds

3. Methodology

In the pilot study shall be carried out that thinned before planting such as Saratovskaya-853, an introduced variety from the Saratov Institute of Russia, was planted at a rate of 3 million sh/ha between May 20 and June 10 in a field. The size of one pad was 3 m^2, and each pad was planted in 9 pads with 3 repetitions. The following observational research was carried out during the study. It includes:

Field germination: Field germination was counted as percentages of sprouts in an 83.3 cm long field in 2 rows of the center of the field when the first leaves were completely uniform in the field.

Growth period: The stage of growth and development of millet was marked as beginning at 15% and as smoothed out at 75%.

Seed yield: The yield was determined by harvesting when the middle part of the millet plant entered the panicle stage.

Biochemical analysis: The quality of millet seeds was determined and expressed as a percentage by the standard of protein (MNS6548:2015) in the Biochemistry Laboratory of the Institute of Plant and Agricultural Sciences.

Quantitative processing: The analysis of variance of research mathematical processing was calculated using R statistical program, and the correlation of factors affecting yield was calculated using the Excell program.

4. Research results

4.1 The effect of the time of planting of field germination

Millet is one of the crops with poor germination shock. According to this feature, small rice can be made when the heat is sufficient. One of the important qualities of the plant's thermal regime is heat supply, which is evaluated by its ability to pass half an hour without changing the temperature. A major determinant of crop yield is field germination. In addition to seed quality and biological characteristics, the yield of the field depends on parameters such as the temperature and moisture regime of the soil of the given year.

Plant seeds grow and absorb moisture depending on their biological characteristics. For example, 45–52% of the weight of the wheat seed is moisture [6, 7] 60–65% of the millet, and 50–55% moisture of the bare oat [8] begins to grow.

According to our research study, in the years of millet field research (2017–2020), it was between 43.1–70.7% in the version of May 20, 46.2–68% in the version of May 30, and 48.9–71.1% in the version of June 10. which means that during the planting period, the climatic parameters of the years are suitable for the growth of warm plants, but the amount of precipitation varies.

In 2017, field protrusions were 43.160.4% in research versions. Out of these, 60.4% of field crops in the scenario of June 30 are 17.3–14.2% higher than other scenarios, respectively. The soil moisture at the time of planting in this period was 19.9 mm at the depth of 0–20 cm, which is 2.6 mm more than the mid-term version and 1.3 mm less than the first time version, but the soil temperature was 30.6°C, which was 15.8–5°C more than the other scenarios. Created the conditions for the increase of protrusions [9, 10].

The differences between the experimental variants were analyzed by factorial analysis of variance in R software. In order to confirm this result, when considering the Tukey T-test in the statistical program R, the version of the time period of June 10 is Pr > 0.000409 from the version of the time period of May 20, and the version of the time period of May 30 is Pr > 0.000797 confirmed to be. The 2018 field germination ranged from 56.0 to 50.7%. As the time of sowing is delayed, the field germination tends to decrease In May-June of this year's planting period, the average daily temperature was 14–20.1°C, which was 0.9–1.1°C higher than the long-term average, and the amount of precipitation fell by 3.7–7.8 mm, which was 16.4–12.3 mm lower than the long-term average, so it was a very dry/arid year.

In the version of May 20, 56% of field germination is 4–5.3% more than in other versions. The temperature of the soil in the first period of planting was 21.8°C, which was 7.1°C lower than the mid-term version and 3.8°C lower than the last version, but it was suitable for plant growth and the amount of soil moisture was 0.9–1.4 mm higher, which caused an increase in field protrusion. Using T-test in R to confirm that the field outliers for the May 20 scenario are better than the other scenarios, the May 20 scenario is significantly different from the May 30 scenario by Pr > 0.022986, and the June 10- Pr > 0.00611 is very different from the time of.

In 2019, the average daily temperature in May-June was 10.7–19.9°C, the sum of active heat was 225-598°C, the heat supply was high enough, and the amount of precipitation was 4.1–48.2 mm which means it was a month lacking in moisture with 17.1–6 mm less precipitation than the average for many years.

The May 20 version is 4.5 to 13.8 higher than the other possible versions, with field germination of 62.7%. The warm soil of this period was suitable for the growth of plants, the amount of soil moisture (0–20 cm depth) at the time of sowing is 19.2 mm, and in the first 10 days of June after sowing, 11.5 mm of precipitation fell, which coincided with the leveling stage of the field germination of the first period, and then created conditions for the increase of protrusions.

An analysis of variance was performed in R to detect differences between the experimental variants, and the variants were different (P > 0.00184). In order to confirm this result, when T-test is also considered in the program, the period of May 20 is higher than the period of May 30 (P > 0.0174614), than the period of June 10 (P > 0.001635), and the period of June 10 It is confirmed that the period is significantly different from the period of May 30 (P > 0.011503) and the field protrusions of the first period of inoculation are better than other versions. According to the 2020 field germination study conducted by us, it was 71.1% in of June 10, which was 0.4–3.1% higher than the scenarios of other periods, respectively. The soil moisture during the planting period is suitable for plant growth. 16.8 mm of precipitation falls in the middle ten days of May and 16.8 mm in the middle ten days of June.

The amount of moisture is sufficient for the plants. However, the average daily temperature was 12.7°C and the soil temperature (V/20–30) at the time of planting was cooler at 15.3–16.8°C, which created the conditions for a decrease in field germination during the first planting period of this year (**Table 1**).

According to our research, the field germination of the May 20 version is 58.1%, which is 2.3–0.6% higher than other versions. The amount of heat at the time of planting increased as the time was delayed, and the amount of soil moisture was more (22 mm) in the scenario of the first time of planting, which affects the field germination. Under the conditions of our country, heat is sufficient for the shoots of millet field germination of the research years, but the size of the shoots varies depending on the amount of soil moisture and the amount of precipitation in that period (**Table 1**).

Indicators	Year	Version		
		V/20	V/30	VI/10
Field germination, %	2017	43.1	46.2	60.4
	2018	56	52	50.7
	2019	62.7	58.2	48.9
	2020	70.7	68	71.1
	average	58.1	56.1	57.8
Soil heat, °C	2017	14.8	25.6	30.6
	2018	21.8	28.9	25.6
	2019	22.8	25.4	27.2
	2020	15.3	16.8	21
	average	18.7	24.2	26.1
Soil moisture, mm	2017	21.2	17.3	19.9
	2018	17.4	16.5	16
	2019	19.2	16	21.1
	2020	30	27.5	23
	average	22	19.3	20
Air temperature, °C	2017	9.1	18.2	22.5
	2018	14	22	19.4
	2019	12.4	13.8	24.4
	2020	13.5	9	13.6
	average	12.3	15.8	20.0

Table 1.
Effect of planting time on field germination of millet (2017–2020).

4.2 Effect of planting time on the survival rate of millet

It is the main indicator of the biological properties of plants. Survival or biological resistance is calculated by comparing the number of plants counted in the field with the number of plants' endurance at harvest. In the conditions of our country, the survival rate expectancy of grain plants is about 65–80%. According to the research of J. Namjilsuren [8], it is 62.1–90.8%of the survival rate of oat and 80–90% in the life of wheat.

According to our research, the average survival rate of millet was 87.1–80.5% in years. According to the life expectancy of the research years 2017–2020, 2017 has a higher life expectancy than other years of 88.3–95.2%. According to the survival rate years, the version of May 20 had an average of 87.1%, which was a 4.2–3.4% survival rate than the other versions (**Figure 1**).

To confirm the results of the planting time variants, analysis of variance in R software showed that the field yield among years was significantly different ($P > 0.00000001870$), but the mean field yield between the variants was not different, $P > 0.64$ (**Table 2**).

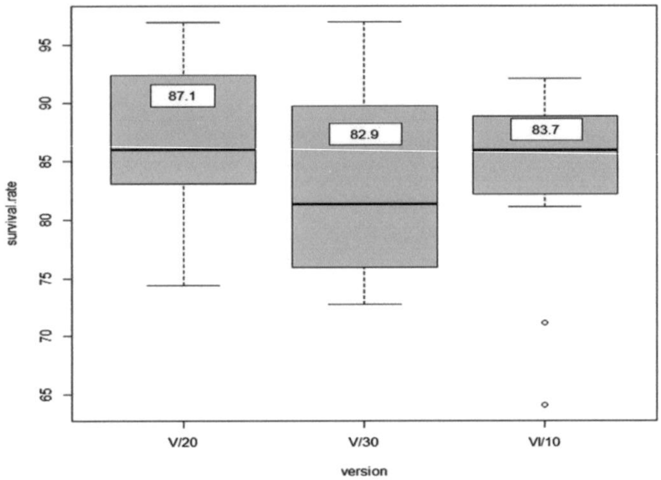

Figure 1.
Effect of millet planting time on the survival rate (2017–2020).

	Df Sum	Sq Mean	Sq F	value	Pr(>F)
Version	2	28	14	0.441	0.6474
Year	1	1827.2	1827.2	57.472	0.0000000187
Version:year	2	275.7	137.9	4.337	0.0222
Residuals	30	953.8	31.8		

Table 2.
Effect of planting time on field germination of millet (2017–2020).

4.3 Effect of planting time on growth period

Millet can grow in different soils, but the yield varies depending on the climate and soil fertility. Setting the right planting time is important not only to increase the yield but also to choose the optimal planting technology according to the soil and climate of the region.

The growing season of any crop may vary due to the biological characteristics of the crop and the soil and climatic conditions of the region. The Saratovskaya-853 variety of millet matures in 84–95 days from the growing period.

Between 1949 and 1957, M. Ölzii carried out research on millet varieties and selected 12 varieties based on the growing period. Also, the Dzungharaa Agricultural Research Institute conducted a comparative study of millet varieties and selected and studied varieties with short growth periods that could yield more under the given research conditions [3].

According to the results of the research, there was a conclusion from many years of research that it is correct to calculate the growing days by including early varieties, reaching 80 days, medium early fruiting 80–90 days, and late fruiting 90–100 days. However, according to the results of research conducted by Researcher J. Serjmaa (1964–1967), the growing days of millet varieties matured in 78–89 days [11].

In addition to factors such as the amount of rainfall and temperature during the planting period, factors such as the quality of the seeds at the time of planting and the characteristics of the variety also affect the time to harvest from the planting of millet. The Saratovskaya-853 variety of millet has 84–95 days from the time of germination to seed maturity. According to the research of J. Serjmaa (1964–1966), it took 26–35 days from cultivation to germination, and 23–28 days from tillering to sprouting of millet varieties [11].

However, according to J. Tsend, the time from germination to threshing of millet varieties was 14–21 days on average, and the development stage of the Saratovskaya –853 variety was uniform in 21 days [6].

Millet starts to be crushed 15–20 days after it starts to be crushed when it has 5–6 leaves. Crushing takes place very well in conditions with an average active heat sum of 270-300°C and an average daily temperature of not less than 17 degrees Celsius.

According to our research, the stages of development of millet from sprouting to crushing were different on May 20, 8–15 days, on May 30, 9–16 days, and on June 10, 9–18 days. Of these, the May 20 version is 12 days earlier than May 30 and 3 days earlier than June 10. In this version of the planting period, the time from the outcrop to the crushing is largely dependent on the amount and distribution of precipitation during this period, which took place in the first 10 days of the end of June and July.

It took 30 days from inoculation to emergence on May 20 and 25 days on May 30. In May, the average daily temperature was 12.8°C, total active heat was 339°C, and precipitation was 23.8 mm. However, in the version planted on June 10, the daily average temperature was 0.8°C higher than the long-term average, 6.95°C higher than the previous month, 22.0 mm higher than the previous month's average, and 8.37 mm higher than the previous month (V). 8.5 days and 5 days earlier than the 2nd-period version, affected by the protrusion.

During the research years (2017–2020), the 5-month daily average temperature of the planting period was 12.85°C, close to the long-term average, the total active heat was 339°C, 34°C more than the long-term average, and the rainfall was 23.8 mm, which was 2.6 mm more than the long-term average had a positive effect.

It takes 32–24 days from the planting of millet to germination. In the scenario of May 20, the germination was uniform for the longest time, but depending on the temperature and moisture of the soil, as the planting time is delayed, the number of days for the protrusions to be uniform is getting shorter. According to our research, the total growing period is 89.5 days for the May 20 version, 85.5 days for the May 30 version, and 80.7 days for the June 10 version. As the growing period is delayed, the growing days become shorter.

The version of the first time of planting was the version of millet that matured in 90 days. On the other hand, the growth period of the June 10 version was 80.7 days, 8.8 days shorter than the May 20 version, and 4.7 days shorter than the May 30 version. Also, during this period, the seed yield is low, the quality is poor, and the crop is formed (**Figure 2**).

4.4 The effect of planting time on yield structural parameters

The of plant yields main factor that determines a particular crop is the yield structure indicator. These vary from crop to crop. Although millet has smaller seeds

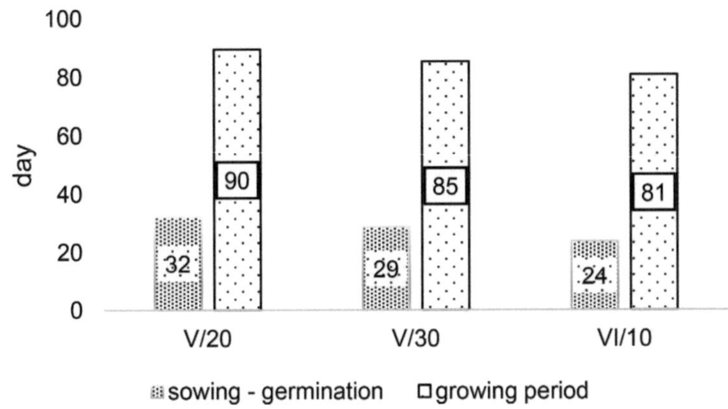

Figure 2.
Effect of millet planting time during the growing period (2017–2020).

Version	Plant height, cm	Length of panicle cm	Number of stems with panicle, sh	Number of seeds per panicle, sh	Seed weight per panicle, g	Weight of 1000 seeds, g
V/20	86.5	14	218	255	1.4	5.4
V/30	79.5	14	192	307	1.6	5.4
VI/10	80.6	12	185	256	1.4	5.3

Table 3.
The effect of planting time on millet yield structural parameters (2017–2020).

compared to other cereal crops, it has similar yield structure parameters. According to J. Namjilsuren's (2003–2004) study, the yield structure measurements made on a total of 62 variety research samples showed that the plant height was 54–104 cm, 3 short samples of 50–80 cm or 3 scores, 81–110 cm medium samples were 19 or 5 scores, and 110- There were 35 samples with a height of 140 cm or 7 points and 5 samples with a height above 140 cm or 9 scores.

The leading indicators of yield structure were plant height, number of seeds per panicle, the weight of 1000 seeds, and length of the panicle. The weight of 1000 seeds of the experimental version is 5.4–5.2 g, the plant height is 86.5–79.5 cm, the number of seeds per panicle is 255.3–306.8 sh/m^2, the weight of seeds per panicle is 1.4–1.6 g/m^2, and panicle length is 14.0–12.3 cm (**Table 3**).

When calculating the relationship between these parameters, the seed yield has a strong direct correlation with the weight of 1000 seeds r = (0.94) and the length of the panicle r = (0.91). However, the yield was moderately correlated with the number of seeds per panicle r = (0.59), and the weight of seeds per panicle r = (0.58). According to the parameters of the structure of the crop, which is highly related to the yield, the weight of 1000 seeds in the version of May 20 was 5.4 g, close to the mid-term version, 0.2 g more than the latest version, and the length of the panicle was 14 cm, which was 0.5 cm more than the first version and 1.7 cm more than the latest version. The plant height was 86.5 cm, which is 7–6 cm higher than in other periods (**Table 2**).

In the analysis of factorial variance in terms of yield structure, P > (0.00008) is very different between plant height variants, (**Table 4**) P > (0.02) difference between

	Df	Sum Sq	Mean Sq	F value	Pr(>F)
Version	2	340.7	170.36	16.89	0.0000089
Residuals	33	332.9	10.09		

Table 4.
Differences between height variants of millet plants, (R program).

time variants in terms of the number of seeds per plant. However, the weight of 1000 seeds was P > (0.06) and the weight of seeds per plant was P > (0.07) and there was no difference between the variants. According to the Tukey T-test, which has a high difference between the versions, there is a significant difference between the May 20 version and the May 30 version P > (0.00001) and the June 10 version P > (0.0001) in terms of plant height., it is confirmed that the May 30 period is better than the May 20 period by P > (0.011) and the June 10 period by P > (0.012).

4.5 Effect of planting time on seed yield

The yield of that crop depends on the precipitation and temperature for 5–6 months of that year. Seed yield varied depending on the growing season of millet. According to research conducted by researcher J. Serjmaa in 1965–1967, the yield of seeds was 16.2–22.2 t/ha [11].

However, according to the research conducted by J. Tsend in 2010–2012, the seed yield of Saratovskaya-853 variety of 17.5 kg/ha [6].

Because comman millet is a crop that matures in succession, it is important to choose the right time to harvest the seeds. Harvesting is done when the upper part of millet ears ripens and starts to turn yellow, because millet/*Panicum miliaceum*. L/seeds ripen in series and also have a lot of spillage. Nesterov.I.M (2018–2020) According to research, the seed yield of millet was 29.6–35.5 t/ha [12].

The 4-year average yield of small rice seeds is 23.7 t/ha in the version planted on May 20, 24.1 t/ha in the version planted on May 30, and 20.9 t/ha in the version planted on June 10. Of these, the version planted on May 30 has a yield of 0.4 t/ha more than the version of May 20, and 3.2 t/ha than the version of June 10 (**Figure 3**).

There was a significant difference in seed yield P > (0.002**) between the years of inoculation, but a small difference P > (0.04249*) between the inoculation time variants. Considering the T-test whether the version of May 30, which has the highest seed yield, is different from other time versions, it is confirmed that the version of this time has a greater difference in yield than the version of June 10 with P > (0.044755*). However, there is no difference between the yields in the version of May 20 P > (0.983759).

We have considered the most beneficial time to sow the millet seed crop to be May 20. During this period, due to the tillage superficial of the soil before planting, the number of weeds in the field was small, and it was not affected by the late spring and early autumn shocks cold. At the germination stage of small rice, the total precipitation was 83.0 mm, which was 15 mm higher than the long-term average, the average daily temperature was 21.2°C, which was 0.3°C higher than the long-term average, and the sum of active heat was 29.4°C, which had a favorable effect on seed yield and quality.

Although the seed yield in the mid-sowing period (May 30) is better than the other options, the main parameters of the field germination, yield structure (plant

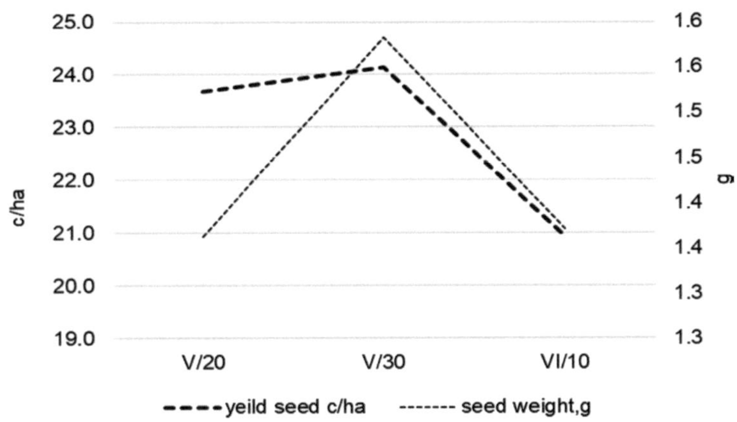

Figure 3.
Effect of planting time on seed yield of millet (2017–2020).

	Df	Sum Sq	Mean Sq	F value	Pr(>F)
Year	1	433.7	433.7	11.1	0.00209
Residuals	34	1328.6	39.1		

Table 5.
Differences between millet seed yields (R program).

height, panicle length), and seed quality are lower than in the first period. Also, due to the climatic characteristics of some years, the sowing period is not suitable due to the fact that after planting in the spring, there were negative consequences such as excessive drying of field germination, drying and death of germination, late seed development, and spot.

On the other hand, the June 10 seed harvest was delayed due to the late sowing and less than the first two periods in yield of parameters, and because it was hit by the autumn frost, it produced seeds with poor germination. Therefore, it is not suitable to plant millet for seed in the central cultivation area after June 10 (**Figure 3** and **Table 5**).

4.6 Effect of planting time on biochemical characteristics

Cereal plants are a source of protein and starch for the world's population. Proteins are unique in that they are not added to other substances when used in human and animal feed. Cereal plants are a source of protein and starch for the world's food, and increasing plant-derived protein is an important issue in meeting the protein needs of the population, so it has become one of the problems facing the world. The protein content varies depending on the type of crop, variety, soil, and climate of the region. Phytochemical composition of grain plants (V.L. Kretovich) wheat seed contains 15% protein, barley 12%, rye 13%, and sorghum 12%. Percent, small rice contains 12%, respectively.

Researcher J. Serjmaa conducted research on small rice varieties from 1964 to 1966, and the three-year average was 751 g of green mass, 16.3% of the weight of seed coat, 6.4% of oil content, 13.3% of acid content, and 48.2% of starch content [13].

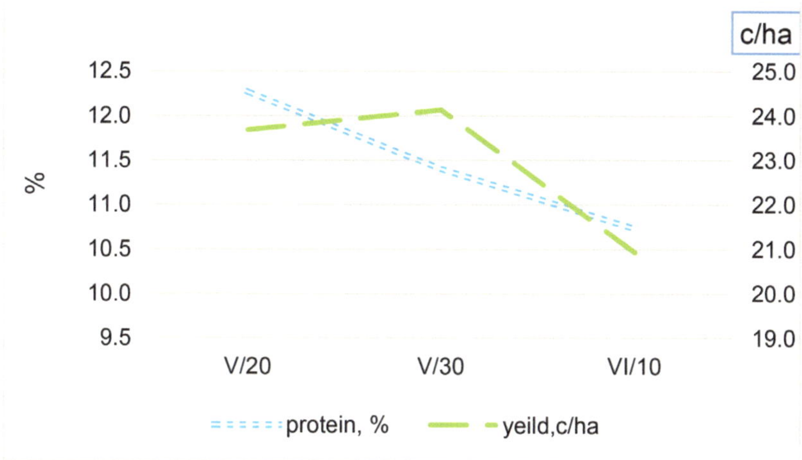

Figure 4.
Effect of planting time on the biochemical quality of millet (2017–2020).

Looking at the chemical composition of plants of the Panicum genus, ordinary millet (*P. miliaceum*. L) contains 12.5 g of protein, 70.4 g of carbohydrates, 2.2 g of fiber, 1.1 g of oil, and 341 kcal of energy contains 60.9-72 g of carbohydrates and 307–341 kcal of energy [14].

According to Lorenz and Dilsaver's [15] research on the milling characteristics, composition, and nutritional quality of millet flour, millet flour had higher ash, fat, and protein content than wheat flour. According to Obilana and Manyasa [16] and Young [17], millet has many nutritional and therapeutic roles and is rich in health-promoting phytochemicals, and is considered a functional food [4].

In our study, the protein content of small rice seeds was between 12.3 and 10.4% in the inoculation variants. On the average of the research years, the protein content of the version planted on May 20 was 12.3%, which was 0.9% higher than the mid-term version and 1.6% more than the late version. When calculating the correlation between the parameters of biochemical characteristics, yield, and protein have a strong correlation of $r = 0.74$ or 74%. The protein content in the seeds of the time variants decreases as the time is delayed, and the quality changes depending on the climatic parameters of the period from germination to maturity, on the average of the years of the study, when the seed condition is mottled and has poor casting (**Figure 4**).

4.7 To compare with the materials of other researchers and to explain the advantages of own research work

Millet is one of the milk-producing crops that have high food and nutritional value, require relatively low moisture, and are resistant to drought.

Research on millet varieties and seed yield was conducted by J. Serjmaa in 1965–1967, and the seed yield was 16.2–22.2 t/ha [13]. And researcher according to J. Tsend's research in 2010–2012, the seed yield of the Saratovskaya-853 variety was 17.5 t/ha [3].

Researcher A.I. According to Baraev's (2015–2017) research in Kazakhstan, the seed yield of millet varieties was between 24.7–33.3 c/ha, which is similar to the results of our study [18, 19].

When millet was planted between V/20–30, the seed yield was 23.7–24.1 t/ha. The results of our research on seed yield are higher than those of previous researchers, which means that in today's climate change, heat supply has a favorable effect on plant growth and is the basis for increasing yield. The cultivation of millet in the conditions of the central cultivation area is an innovative and superior study suitable for the conditions of our country, which is the main cause of climate change, the reduction of crop yield, the production of healthy and safe food, and the production of healthy and safe food due to climate change.

5. Conclusion

1. In millet field germination the version planted on May 20 had 58.1% of shoots, which was 2–0.3% more than the other time versions. The amount of soil moisture in this scenario was 2–1.7 mm higher than that of the other scenarios, which created the conditions for the increase of the field germination, and the moisture field germination at the time of planting has a correlation of $r = 0.7$.

2. In the May 20 time version, the growing period was 90 days, which was the most suitable time for seed ripening, which was 5–9 days longer than the other versions.

3. It is confirmed that the version of May 20 is 12.3% of the protein content in the seeds and is better than other versions in terms of quality content. Nutrient quality indicators of millet decrease as the time is delayed, which is largely dependent on the condition of the seed.

4. According to the 24.1 t/ha in the version planted on May 30 although seed yield was 0.4–3.2 t/ha higher than the other versions, it did not reach the version of May 20 in terms of seed quality, field germination, and yield structure parameters.

5. May 20 is considered the most suitable time for sowing millet seeds. The seed yield of this version is 23.7 t/ha, which is no different from the version of May 30 ($P > (0.983759)$), and 11.8% higher quality and uniform seeds than the version of June 10.

Author details

Khishigbuyan Turbat[1*], Gungaanyam Galkhvv[2] and Namjilsuren Jamiyan[2]

1 Mongalion University of Life Sciences, Darkhan-uul Province, Mongolia

2 Plant Science and Agricultural Institute, Mongolia

*Address all correspondence to: bankuhish@gmail.com

References

[1] Gungaanyam G, Khishigbuyan T. Warming-tolerant Native Crops. Article. Darkhan-uul province; 2015

[2] Available from: https://www.fao.org/millets-2023/en

[3] Tsagaanshuher G. Plant Farming. Textbook. Darkhan-uul province; 2001

[4] Habiyaremye C, Mataguinan JB. Proso millet (*Panicum miliaceum* L.) and its potential for cultivation in the Pacific Northwest, U.S.: A review. Frontiers in Plant Science. Jan 2017;**2**. Available from: https://www.frontiersin.org/articles/10.3389/fpls.2016.01961/full

[5] Habiyaremye C. Proso Millet (*Panicum miliaceum* L.) and It's Potential for Cultivation in the Pacific Northwest, USA: A Review. 2017. DOI: 10.3389/fpls.2016.01961. Available from: https://www.frontiersin.org/articles/10.3389/fpls.2016.01961/full

[6] Tsend J. Research on Millet Varieties and the Effect of Planting Time in the Central Agricultural Region [Master's thesis]. Darkhan-uul province; 2012

[7] Nambar J. The result of the study of agrotechnics for growing winter wheat in the conditions of soil compaction technology [Dissertation for the Ph.D]. 2002. pp. 42-43

[8] Namjilsuren J. Selection of bare-seeded oat varieties in the conditions of the central cultivation area, results of research on planting time and seed norms [Dissertation for the Ph.D]. 2002. pp. 95-106

[9] Khishigbuyan T. The influence of planting time on the green mass yield and quality millet. In: Sustainable Development of Environment and Agriculture. Scientific Conference Articles. Research Articles. Mongolia: Darkhan-uul provenci; 2019. pp. 101-104. ISBN: 978-99973-935-7-9

[10] Khishigbuyan T, Gungaanyam G, Namjilsuren J. The influence of planting time of the seed yield and quality millet. Mongolian Journal of Agricultural Sciences. 2018;**25**(03):127-131. DOI: 10.5564/mas.v25i03.1181

[11] Serjmaa J. Millet. Mongolia: Ulaanbaatar; 1977

[12] Nesterov IM. Influence of sowing time on the yield of millet grain in the conditions of the north-eastern part of Belarus. 2018-2020. Удк 633.17:631.559(476-18). Available from: https://cyberleninka.ru/

[13] Aliscioni SS, Giussani LM, Zuloaga FO, Kellogg EA. A molecular phylogeny of (Poaceae: Paniceae): Tests of monophyly and phylogenetic placement within the Panicoideae. American Journal of Botany. 2003;**90**(5):796

[14] Tomé D, Bos C. Lysine requirement through the human life cycle. Journal of Nutrition. 2007;**137**(6 Suppl 2):1642S-1645S. DOI: 10.3732/ajb.90.5.796

[15] Lorenz K, Dilsaver W. Proso millets. Mil. 1980. Available from: https://www.cerealsgrains.org/

[16] Thapliyal V, Singh K, Millet F. Potential millet for food security and power house of nutrient. International Journal of Research in Agriculture and Forestry. Feb 2015;**2**(2):22. ISSN 2394-5907 (Print) & ISSN 2394-5915 (Online)

[17] Amadou I, Gounga ME, Le G-W. Food science and nutrition millets: Nutritional composition, some health benefits and processing - A review. Emirates Journal of Food and Agriculture. 2013;**25**(7):501. DOI: 10.9755/ejfa.v25i7.12045. Available from: http://www.ejfa.info/

[18] Baraeva AI. Влияние фактор среды на прочатность зерна и зеленой масси проса в severnom kazakhstane. Available from: https://kazatu.edu.kz/

[19] Anokhina TA. O tselesoobraznostnosti oslojannosti prosa v khaslitte trehovoy kultury. No. 1. – S. 6. 5. 2004

Chapter 7

Seeds of Resilience: Physiology and Mechanisms of Hardseededness

Sıtkı Ermis, Eren Özden and Ertan Yildirim

Abstract

Physical dormancy, also known as hardseededness or seed coat impermeability, is a condition that occurs when a seed's coat becomes impermeable, preventing the entry of water, gases, and other external factors. This impermeability serves as a protective mechanism, delaying germination until suitable conditions are met. Factors influencing hard seed formation fall into two categories: internal and external. Internal factors pertain to plant-specific traits, such as species and seed morphology. Genetic variations and seed coat characteristics play a role in shaping hard seed formation. External factors, based on environmental conditions, also influence seed development. Soil nutrient availability, water supply, humidity, temperature, and light conditions impact seed coat permeability and germination. Additionally, the timing of seed maturity, drying, and storage conditions can contribute to hard seed formation. The interplay of these factors determines a plant's tendency to produce hard seeds. Overcoming dormancy caused by seed coat impermeability involves various methods, including physical, chemical, and mechanical approaches. These methods enhance water and gas permeability, facilitating germination. The choice of method depends on seed characteristics and desired outcomes in breaking dormancy. This section emphasizes the impact of hardseededness on seed quality and the application of methods to enhance germination, underscoring its significance in seed science.

Keywords: hardseededness, seed coat, germination, breaking hardseededness, factors responsible for physical dormancy

1. Introduction

Seeds, a remarkable wonder of nature, are the cornerstone of plant life. However, they are not just the beginning of a plant's life; they also encompass many valuable lessons about the resilience of the natural world. This phenomenon, known as "Seed Resilience," represents the fascinating story of how seeds possess the ability to protect themselves and adapt to environmental challenges.

Seed dormancy is a natural process in which seeds delay their germination, even when the environment is suitable for it. During the dormancy period, seeds typically remain in a dormant state, often protected by mechanisms that prevent early germination. Dormancy permits seeds to germinate when favorable conditions arise, and the diverse levels of dormancy within a population of seeds contribute to a gradual and staggered germination process over an extended period. This chronological variation in germination timing plays a critical role in ensuring the species' survival, especially in demanding environmental circumstances [1].

While seed dormancy is generally viewed as an unfavorable trait in agriculture, where the primary objective is to promote rapid seed germination and growth, however, there are situations where seed dormancy can offer significant advantages, particularly during the seed development stage. This advantage is particularly noteworthy in the case of cereal crops. Cereals possess a dormancy mechanism that prevents germination while the grains are still attached to the parent plant's ear. This mechanism acts as a crucial safeguard, especially when there is a period of rainfall during harvest (known as preharvest sprouting), as it prevents premature germination of cereal crops and helps avoid substantial losses in the agricultural sector [2].

Conversely, weed seeds often maintain their inherent dormancy mechanisms as they mature. This allows some weed seeds to persist in the soil for many years, patiently awaiting the right conditions for germination. This poses a threat to crop cultivation, as these seeds can rapidly multiply when favorable conditions finally occur. In summary, seed dormancy is a vital consideration in agriculture and crop cultivation, playing a significant role in the survival strategies of plant species [3].

Seed dormancy is not a single, straightforward trait; rather, it is a complex phenomenon influenced by various factors. Its complexity arises from the intricate interactions between these factors, and it is shaped by multiple elements. Among the numerous factors affecting seed dormancy, hardseededness stands out prominently [4]. Also known as physical dormancy or seed coat impermeability, hardseededness is a fascinating botanical phenomenon that has intrigued researchers and environmental scientists for a long time.

This distinctive feature, present in many plant species, serves as a crucial adaptation strategy. It allows seeds to postpone germination until favorable conditions emerge, bolstering the seed's ability to withstand challenging environmental conditions [5]. It is a phenomenon that is an integral part of a seed's life cycle and is a product of the natural selection process. Unlike seeds that readily sprout under suitable conditions, hard seeds boast a robust protective seed coat, serving as a formidable barrier that prevents premature germination, ensuring the seed's survival until optimal conditions prevail.

In this chapter, we focus on defining seed resilience, understanding why it is important, and exploring the different types of hard-seededness. Additionally, examines the techniques used to overcome physical dormancy and discusses the factors responsible for the formation and maintenance of this dormancy, aiming to explore various aspects of seed resilience.

2. Seeds of resilience

2.1 Definition of physical dormancy

Physical dormancy trait is an inherited characteristic that originates from both the outer and inner structures of the seed during its development [6]. Physical dormancy is recognized in 18 plant families that span a wide range of taxonomic groups and geographical locations worldwide [7–10]. Studies have suggested that physical dormancy is often inherited as a Mendelian trait, with specific genes controlling the development of impermeable seed coats. The heritability of this trait varies among plant species, indicating a complex genetic basis influenced by both dominant and recessive alleles [11, 12]. When it comes to the genetic basis of hardseededness, two types of alleles come into play: dominant and recessive. Dominant alleles are the ones responsible for the formation of impermeable seed coats. Even a single dominant allele for hardseededness is sufficient to yield seeds adorned with impermeable seed coats. These robust seed coats serve as formidable shields, thwarting the intrusion of external elements [13]. On the other hand, recessive alleles lead to the formation of permeable seed coats. However, for a plant to produce permeable seeds, it must carry two recessive alleles, one from each parent. In other words, if both parent plants contribute a recessive allele for hardseededness, the resulting seeds will have permeable seed coats. This genetic interplay between dominant and recessive alleles determines whether a seed's coat will be impermeable or permeable, ultimately influencing its ability to withstand external environmental conditions [14].

The degree of physical dormancy is influenced not only by genetics but also by several environmental factors. The primary factor influencing hardseededness in seeds is their moisture content. The level of moisture within seeds plays a critical role in determining their ability to germinate. Seeds with low moisture content frequently display hardseededness because the limited availability of moisture impedes the essential biochemical processes required for germination. In simpler terms, when seeds lack sufficient moisture, they tend to become hard and resistant to germination. It is widely accepted that hardseededness is typically a permanent trait. Once a seed coat becomes impermeable, it is considered unlikely to return to a permeable state unless there is damage to the seed coat or a specific structure within it, known as the "water gap" opens to enable water to reach the internal structures and rehydrate them [5, 15]. Many studies on seed development have demonstrated that the transition from a permeable to an impermeable seed coat coincides with the decrease in moisture content during the maturation drying phase of seed development [16–18].

Temperature is also undeniably the primary environmental factor influencing physical dormancy in seeds as a key element in synchronizing plant growth with changes in climate conditions. Many studies have investigated the relationship between temperature during seed maturation and the permeability of the resulting seeds [19–21]. These studies have consistently shown that seeds that mature at higher temperatures tend to produce a higher proportion of impermeable seeds compared to those that mature at lower temperatures. It is important to note that measuring temperature throughout the seed maturation period can be challenging, and seasonal or location-specific temperature variations around the parent plant may also play a role in determining the permeability of seeds. For example, seeds maturing at higher temperatures are likely to experience more water loss, leading to a higher proportion of seeds with physical dormancy.

2.2 Physical dormancy and germination

Determining when genetic and physiological differences emerge is a challenging endeavor, primarily because seed dormancy is intricately regulated at different stages of development, often influenced by environmental factors [22]. A dormant seed is essentially incapable of initiating germination, even when presented with favorable environmental conditions and a suitable habitat. This incapacity can be attributed to various factors, such as non-viability, the absence of an embryo, or the presence of dormancy. The impermeability of seed coats in physically dormant seeds is the result of complex structural and chemical adaptations (**Figure 1**).

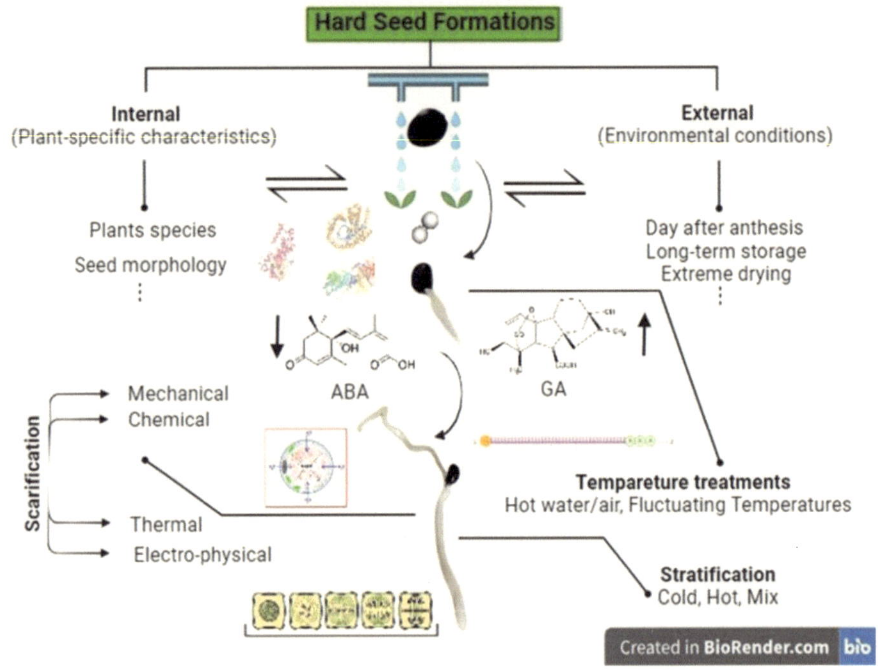

Figure 1.
Physical dormancy in hardseed formulations.

Germination encompasses a series of events that start with the absorption of water by a resting, desiccated seed and concludes with the elongation of the embryonic axis, ultimately resulting in the emergence of the radicle [23]. Water uptake by a seed occurs in three distinct phases. The first phase, known as Phase I, is characterized by a rapid and initial uptake of water. This phase typically occurs within 1–8 h. Water uptake occurs in all seeds, whether dormant or non-dormant, and even in non-viable seeds. In fact, non-viable seeds tend to absorb more water compared to viable seeds because the turgor pressure in the cells of viable seeds restricts excessive water uptake. Phase II, also referred to as the lag phase, marks a stage where water uptake becomes limited. However, during this phase, essential enzymes like amylase, endoxylanase, and phytase are produced. These enzymes are responsible for synthesizing new proteins required for germination. Additionally, the conversion of stored materials for germination, mRNA synthesis, and the initiation of energy production through sugar metabolism degradation also begin during this phase. Consequently, this phase facilitates the transfer of nutrients from storage areas in the endosperm to the growth points. Phase II can span from several hours to a few days and typically concludes when the radicle emerges from the seed coat. The third phase, Phase III, signifies the first visible sign of germination, with the emergence of the radicle. This emergence is primarily a result of cell enlargement rather than cell division. Shortly thereafter, cell division occurs at the tip of the radicle, initiating its elongation [24–28].

Physical germination is an event that occurs through a two-stage process involving the rupturing of the seed coat. The first stage involves the rupture of the seed's outer covering, known as the testa. This protective layer shields the embryo within the seed, and its removal is necessary to initiate germination. The second stage involves the rupture of an inner layer called the endosperm. This inner layer serves as a storage site for nutrients within the seed and provides the essential nutrients for the initial growth of the embryo [22, 29].

Germination is completed, specifically the emergence of the radicle (the embryonic root) becomes visible. This marks a clear indicator of seed vitality and the beginning of germination. However, in seeds with physical dormancy, the seed coat is tough, which can limit water uptake. As a result, germination may either not occur at all or be restricted when water uptake is insufficient. Physical dormancy is an evolutionary strategy that prevents seeds from germinating until environmental conditions are favorable, and water uptake is a crucial component of this process. Most of the time, the seed coat limits germination by either preventing water and/or oxygen from passing through or by resisting the emergence of the radicle with its mechanical properties [30]. In this process, the seed coat serves as a vital organ involved in nourishing the embryo and subsequently safeguarding it against harmful environmental factors [31, 32].

Dormancy caused by seed coat impermeability can be overcome through various methods, including physical, chemical, and mechanical approaches. These methods aim to enhance water and gas permeability in hard seeds, facilitating moisture absorption and initiating germination. The choice of method depends on specific seed characteristics and desired outcomes in breaking dormancy.

3. Methods of breaking physical dormancy

Breaking physical dormancy involves various methods to weaken or remove this barrier, allowing water and oxygen to penetrate the seed and initiate germination. Physical dormancy may require different scarification methods involving either physical or chemical agents to overcome seed dormancy.

3.1 Effect of seed scarification

Seed scarification is a crucial technique used in agriculture and horticulture to promote germination by weakening the seed coat. This process involves various methods to facilitate water absorption and oxygen penetration into the seed, ultimately kickstarting the germination process. Various methods can be employed for seed scarification, including chemical, thermal, electro-physical, and mechanical approaches [33].

The predominant form of scarification is mechanical scarification, wherein the testa, or seed coat, undergoes physical manipulation to enable the ingress of moisture and air. Various approaches are utilized in this process, encompassing the abrasion of seed coats with tools such as metal files, sandpaper, knives for nicking, gentle hammering for cracking, or any method that weakens or opens the seed coat [34–39]. This technique finds extensive application within the realms of horticulture and agriculture, serving as a fundamental method to facilitate the germination of seeds encased in rigid or impermeable seed coats. Its primary function lies in enabling seeds to surmount physical dormancy, thereby instigating the pivotal processes of water absorption and germination. Mechanical scarification stands as a pragmatic and efficacious method, widely adopted to unlock the growth potential of numerous plant species reliant on this mechanism for breaking their dormancy and flourishing within diverse ecological contexts.

Chemical scarification is a seed treatment method that employs various chemical substances, including potent acids such as H_2SO_4 (sulfuric acid), HCl (hydrochloric acid), $NaHClO_3$ (sodium hypochlorite), and hydrogen peroxide (H_2O_2). Also, organic solvents such as alcohol and acetone are used to eliminate seed dormancy. These chemicals have been the subject of extensive research, revealing their capability to effectively disrupt and overcome the dormancy mechanisms present in specific types of seeds. Through the application of these chemicals, the hard or impermeable seed coats of certain plant species can be altered or weakened, allowing for enhanced water penetration and subsequent germination. In recent studies, chemical scarification with H_2SO_4 has been particularly effective in enhancing seed germination and seedling growth in various plant species. For instance, in okra [40, 41], strelitzia [42], muscari [43], and raspberry seeds [44], H_2SO_4 treatment has been found to break physical dormancy barriers and promote successful germination. Breaking physical dormancy can also take place as seeds pass through an animal's digestive system. Unlike conventional approaches to dormancy relief, exposure to hydrochloric acid (HCl) within the digestive tract gradually erodes the impermeable seed layers [45, 46]. As well as the scarification treatment involving HCl acid is recognized for its high effectiveness in promoting germination for species with hard seed coats. This method involves the use of HCl to weaken or modify the tough outer seed coat, allowing water to penetrate and initiate germination more easily. Various methods have been employed to overcome physical dormancy, and they have proven effective in breaking seed dormancy and enhancing germination success in a range of plants, including guava [47], sunflower seeds [48], Indian siris [49], as well as certain forest tree seeds [50]. Sodium hypochlorite scarification is one such method that has proven highly efficient in breaking physical seed dormancy. Recent studies [51, 52] have reported significant improvements in seed germination, notably in *Andrographis paniculata* and *Vanilla planifolia*, following $NaHClO_3$ scarifications. In some cases, this scarification is combined with ethanol pre-treatment, with ethanol concentrations ranging from 50 to 96% [53, 54], further enhancing the dormancy breakage process. Hydrogen peroxide has also shown promise in releasing physical seed dormancy; for instance, Kindinger [55] reported encouraging results when soaking *Tripsacum dactyloides* (eastern gamagrass) seeds in a 30% hydrogen peroxide solution for 2 hours. Additionally, organic solvents like alcohol (typically ethanol) and acetone are employed for seed scarification, which is aimed at breaking dormancy. This process involves soaking seeds in the chosen organic solvent for a specific duration, which varies based on the seed type and its dormancy characteristics. Prior studies have consistently demonstrated the highest germination percentages using acetone and ethanol in various seeds such as tomato, lettuce, sunflower, and grass [56–58]. These methods play a crucial role in improving germination rates and enhancing plant propagation.

3.2 Effect of temperature treatment

The impacts of wet heat, dry heat, and the alternating use of wet heat and ice water on the alleviation of seed dormancy play crucial roles in the process of seed germination and propagation.

3.2.1 Hot water/air treatment

The process of hot water or air treatment entails immersing seeds in hot water or exposing them to high-temperature air. This procedure serves to soften the seed coat, eliminate any wax or grease present, and facilitate the penetration of water into the seed under natural conditions. These treatments hold the potential to physically disrupt or soften the macrosclerid layer, with a specific focus on the strophiolar plug. This may lead to the creation of cracks, a softer plug, or alterations that enhance water absorption and other germination-related processes [59, 60].

The presence of cracks or fissures in the seed coat plays a critical role in initiating the germination process. These openings act as conduits through which water can enter the seed. Once water infiltrates the seed, it sets off a series of vital biochemical processes necessary for germination. This mechanism has been extensively documented in the research conducted by Baskin and Baskin in [61]. Furthermore, the use of hot air or water as a treatment for seed germination can have additional consequences beyond facilitating water entry. It may expose the embryo to thermal shock, characterized by a sudden temperature change that can affect its metabolic processes. The impact of this thermal shock can vary among different species, potentially either promoting or inhibiting germination, contingent on a species' adaptability to temperature fluctuations [62]. The efficacy of dry heat treatment, on the other hand, is contingent on the extent and duration of exposure. Therefore, determining the optimal combination of temperature and duration that yields a high germination rate necessitates empirical research tailored to each specific species [63].

This method is also proficiently employed to break seed dormancy in rice [64], okra [65, 66], black mimosa (*Mimosa bimucronata*) [67], and sponge gourd [68], resulting in higher seed germination rates compared to control groups. In a distinct study conducted by Taghizadeh and Sajadi in [69], the most effective seed germination treatment for Spanish broom (*Spartium junceum* L.) was introduced. This involved immersing the seeds in boiling water, subsequently reducing the mean germination time.

Another method employed to overcome hard seed dormancy is a combination of temperature and moisture, often referred to as after-ripening. It is utilized to break hard seed dormancy, particularly in seeds with physical or physiological dormancy. It involves subjecting seeds to specific temperature and moisture conditions, which trigger a series of biochemical and physiological changes that ultimately promote germination [70]. After-ripening takes place in a broad spectrum of warm and dry environmental settings, and the particular combinations of temperature and relative humidity (RH) significantly impact the after-ripening rate of different species. Numerous investigations into after-ripening have been conducted, during which the moisture content (MC) of seeds was assessed, and the seeds were subsequently stored at defined temperatures while being observed until their dormancy was disrupted. Heat and moisture combination treatments were discovered to be effective in overcoming seed dormancy in Malvaceae species, as reported by Demir [66] and Ellis et al. [71]. Furthermore, physical dormancy in seeds of *Ipomoea* spp. [72], *Geoffroea decorticans* [73], and *Pittocaulon praecox* [74] are broken when exposed to heat temperatures. During these studies, seeds were stored at specific temperatures, and the impact of these temperatures on the after-ripening process was observed. The seeds were regularly monitored at these temperatures until their dormancy was broken.

The well-documented role of fire in aiding the germination of species naturally found in fire-prone regions is significant. The process of vegetation combustion, as seen in forest fires, releases a variety of factors that actively promote seed germination and facilitate the breaking of seed coats, allowing water to penetrate for germination. Due to the heat shock from the fire, the specialized, heat-sensitive tissues within the seed coat are either broken or displaced, leading to the creation of 'water gaps' [17, 75, 76]. This allows the seed to become capable of absorbing water and oxygen, which in turn aids in the ongoing process of germination. Several studies have presented compelling evidence that smoke and its components can initiate seed germination and promote growth after germination [77–80]. The nature and extent of the temperature required to break seed dormancy and the response to temperature can vary depending on several factors. These factors encompass the characteristics of the seed coat, such as the thickness and structure of the cuticle, epidermis, palisade, parenchyma, and the type of water gap [81]. Seed size is also a significant consideration, as smaller seeds tend to exhibit higher temperature tolerance and lower temperature thresholds [82]. Additionally, it has been observed that the age of the seed can influence its response to temperature [83]. Finally, whether the temperature conditions are moist or dry can also shape the seed's response [84].

3.2.2 Exposing seeds to fluctuating temperatures and lights

Physiological dormancy can be alleviated not only through the process of after-ripening in dry storage conditions but also by subjecting seeds to variable temperature conditions. The specific temperature exposure required for this process can vary depending on the needs and germination characteristics of different plant species. The choice of temperature ranges and the duration of exposure may vary based on the species of seeds and their sources. In 1916 [85], Harrington's research demonstrated that seeds of Trifolium and Melilotus underwent a softening process when exposed to temperature fluctuations, alternating between 10°C or lower and 20°C or higher. This softening effect was significantly enhanced when the seeds had previously experienced cooler temperatures below 10°C. In another study, exposure to specific temperature conditions, such as warm stratification (WS) and cold stratification (CS), increased the germination rate to over 80% [86]. However, it has been shown that seed germination behavior is often associated with factors such as habitat, seed mass, and life cycle type. For instance, many plants in moist habitats like wetlands tend to have their seed germination promoted by temperature fluctuations, while forest plants typically do not respond positively to temperature fluctuations [87]. Fenner and Thompson [88] suggested that small-seeded plant species may be more stimulated by temperature fluctuations compared to larger-seeded species.

While all seeds require water, oxygen, and appropriate temperature conditions for germination, some species also need light to germinate. It has been observed that light filters through the seed coat, and specific wavelengths of light can penetrate the embryo [89]. Generally, larger-seeded species do not require light during germination because they have enough nutrients to grow in the dark. On the other hand, smaller-seeded species need light to germinate. In fact, these seeds remain dormant even if they have absorbed water when planted at a depth where they cannot access light. Light requirement is a dormancy factor and may not prevent germination in vegetable species except under extreme conditions. However, wild species in nature and tree seeds in forests generally have a greater requirement for light. The effect of light on seeds varies depending on genotype and environmental conditions during seed maturation, dormancy breaking, and germination [23]. Light and gibberellins (GA) are also two key factors that can help overcome this type of dormancy. It has been suggested that as a photoreceptor, phytohormone can contribute to the regulation of light in initiating or terminating the processes of seed dormancy [90]. Some seeds have light-sensitive pigments in their seed coats, and exposure to specific wavelengths of light (usually red or far-red light) can trigger biochemical changes in the seed coat. This can lead to the degradation of inhibitory compounds and the softening of the seed coat, allowing water and oxygen to penetrate the seed, which is essential for germination [91–93]. However, the mechanism by which light promotes or inhibits seed germination has not yet been fully understood [94].

Gibberellins, a class of plant hormones, play a crucial role in regulating various growth and developmental processes within plants, including the germination of seeds. One of the primary functions of gibberellins in seed germination is their ability to stimulate the production of specific enzymes, such as α-amylase [95]. These enzymes are responsible for breaking down stored starches present in the endosperm of the seed, which provides the energy for the growth of roots and shoots [96]. During the early stages of seed germination, the embryo relies on these stored starches as a source of energy and essential nutrients. The breakdown of starches by alpha-amylase leads to the formation of soluble sugars, which serve as a vital energy source for the developing embryo [97]. This energy is necessary to support the growth and emergence of the embryonic shoot and root from within the seed. In addition to their role in providing energy, gibberellins also contribute to alleviating dormancy in seeds. Dormancy is a state of inhibited growth and development that some seeds enter to ensure their survival until favorable conditions for germination are met. Gibberellins promote dormancy release, effectively signaling to the seed that it is time to initiate the germination process. Moreover, gibberellins can influence the physical properties of the seed coat. They have the capacity to soften the seed coat, making it more permeable to water and gases. The function of gibberellins in dormancy alleviation is combined with the necessity for light [98]. This softening of the seed coat is particularly important for the emerging radicle, which is the embryonic root of the plant. The softening of the seed coat facilitates the penetration of the radicle through the seed coat, allowing it to emerge more easily and initiate the germination process.

The increase in GA levels occurs with the release of seed dormancy, for example, during imbibition and stratification. In some dormant seeds, treatment with exogenous GA can take advantage of this opportunity to break dormancy, thereby facilitating faster seedling production [99]. Many previous studies have utilized GA to facilitate the breakage of dormancy in various plant species, including *Solanum torvum* [100], Chinese ryegrass (*Leymus chinensis*) [101], *Amsonia elliptica* [99], *Cinnamomum migao* [102], and Araticum (*Annona sylvatica*) [103]. However, it is important to note that the responses to GA treatment have displayed variability depending on factors such as the specific plant species, dormancy type, GA forms, concentration, and duration of treatment.

4. Bibliometric analysis

To gain a comprehensive insight into the research landscape concerning physical dormancy and hardseededness within the agricultural domain, a thorough bibliometric analysis was undertaken (**Figure 2**).

In the bibliometric analysis conducted based on the content indexed in Web of Science, all agricultural fields were selected with the keywords "Physical Dormancy" and "Hardseededness" 675 results were obtained in English documents. According to years, the oldest being 1990 and the newest being 2023, accessed 629 articles, 25 proceeding papers, 21 review articles, 7 early access, 4 editorial materials, 3 book chapters, 2 meeting abstracts and 1 research note from 31 different disciplines. As seen in **Figure 2**, it is divided into 6 different clusters according to the author's keywords, and

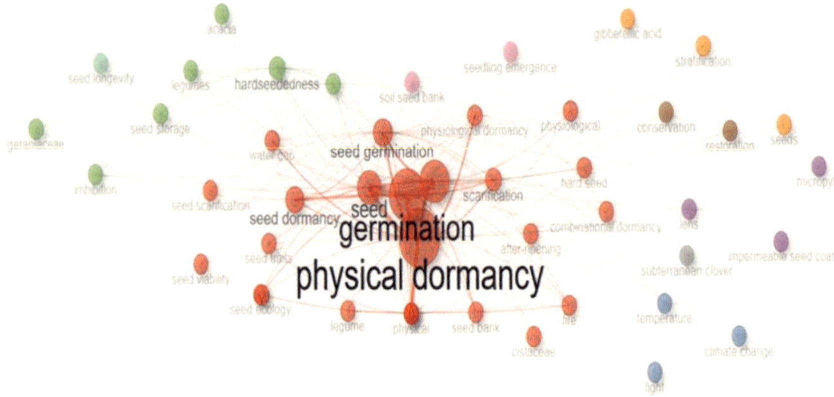

Figure 2.
Visual representation of keywords based on co-word (co-occurrence) using RStudio.

in the red cluster, the term physical dormancy is included in 167 works, germination is 136, seed dormancy is 113, dormancy is 103, seed dormancy is 60 works and hardseededness is 49 works. In physical dormancy studies, 1836 different authors were found, 24.74% of which were multinational studies. Considering the topicality of the research topic, it was observed that the document average age was 10.1 years, and the average number of citations per document was 17.07.

5. Conclusions

In summary, seeds represent a remarkable natural wonder, serving as the bedrock of plant life and imparting valuable lessons about the resilience of the natural world, a concept known as "Seed Resilience." Seed dormancy, a natural process that delays germination even in optimal conditions, plays a vital role in ensuring the survival of plant species by enabling staggered germination to adapt to challenging environments. Physical dormancy, or hardseededness, is a specific type of dormancy characterized by an impermeable seed coat, acting as a protective shield to prevent premature germination and safeguard the seed until favorable germination conditions prevail. The factors influencing the formation of hard seeds can be categorized into internal (plant-specific) and external (environmental) factors, encompassing genetic differences among plant species, morphological seed features, and environmental conditions like soil quality, temperature, humidity, and light exposure.

The breaking of dormancy in hard seeds can be accomplished through various methods, including mechanical abrasion, chemical treatments, temperature fluctuations, and cycles of drying and rehydration, all aimed at improving water and gas permeability for successful germination. Understanding physical dormancy is of utmost importance for researchers, ecologists, and horticulturists, as it significantly impacts seed germination timing and success, with broad implications for plant populations, crop production, and ecosystem restoration efforts.

Furthermore, the role of light in regulating dormancy and germination remains a topic of ongoing debate. Light has been recognized both as a stimulant for germination and as a dormancy terminator, contingent upon the perspective taken. It is considered the final step in the dormancy-breaking process, allowing seeds to germinate even in darkness. Particularly, red light, mediated through phytochrome, can reverse dormancy and promote germination. In seeds with coat dormancy, both light and gibberellins (GA) are believed to release dormancy and facilitate germination.

In the field of bibliometrics, an extensive body of research in this area is evident, with numerous authors contributing to the understanding of physical dormancy, including a significant portion of multinational studies. The research is current and well-cited, underscoring the significance of this topic. In conclusion, the intricate interplay of factors that influence seed dormancy and germination is a complex subject under ongoing research and debate. A comprehensive understanding of these mechanisms is vital for effective seed management and ecosystem restoration practices.

Conflict of interest

The authors declare no conflict of interest.

Author details

Sıtkı Ermis[1], Eren Özden[2,3] and Ertan Yıldırım[4*]

1 Faculty of Agriculture, Department of Horticulture, Osmangazi University, Eskişehir, Türkiye

2 Faculty of Agriculture, Department of Horticulture, Iğdır University, Iğdır, Türkiye

3 Faculty of Agriculture, Department of Horticulture and Agronomy, Kyrgyz-Turkish Manas University, Bishkek, Kyrgyzstan

4 Faculty of Agriculture, Department of Horticulture, Atatürk University, Erzurum, Türkiye

*Address all correspondence to: ertanyil@atauni.edu.tr

References

[1] Bewley JK, Nonagaki H. Seed maturation and germination. In: Roitberg BD, editor. Book of Reference Module in Life Sciences. Matsudo, Japan: Elseiver; 2017. DOI: 10.1016/B978-0-12-809633-8.05092-5

[2] Bewley JD, Bradford KJ, Hilhorst HWM, Nonogaki H. Seeds: Physiology of Development, Germination and Dormancy. New York, NY, USA: Springer; 2013. p. 392. DOI: 10.1007/978-1-4614-4693-4

[3] Qasem JR. Weed seed dormancy: The ecophysiology and survival strategies. In: Carlos J-LC, editor. Book of Seed Dormancy and Germination. London, UK: IntechOpen; 2019. pp. 1-36. DOI: 10.5772/intechopen.88015

[4] Graeber KAI, Nakabayashi K, Miatton E, Leubner-Metzger G, Soppe WJ. Molecular mechanisms of seed dormancy. Plant, Cell and Environment. 2012;**35**(10):1769-1786. DOI: 10.1111/j.1365-3040.2012.02542.x

[5] Baskin CC, Baskin JM, Li X. Taxonomy, anatomy and evolution of physical dormancy in seeds. Plant Species Biology. 2000;**155**:139-152. DOI: 10.1046/j.1442-1984.2000.00034.x

[6] Hudson AR, Ayre DJ, Ooi MK. Physical dormancy in a changing climate. Seed Science Research. 2015;**25**(2):66-81. DOI: 10.1017/S0960258514000403

[7] Baskin CC. Breaking physical dormancy in seeds: Focusing on the lens. New Phytologist. 2003;**158**:229-232. DOI: 10.1046/j.1469-8137.2003.00751.x

[8] Baskin JM, Baskin C, Dixon KW. Physical dormancy in the endemic Australian genus Stylobasium, a first report for the family Surianaceae (Fabales). Seed Science Research. 2006;**16**:229-232. DOI: 10.1079/SSR2006248

[9] Jayasuriya K, Baskin JM, Baskin CC. Sensitivity cycling and its ecological role in seeds with physical dormancy. Seed Science Research. 2009;**19**:3-13. DOI: 10.1017/S096025850818730X

[10] Gama-Arachchige NS, Baskin JM, Geneve RL, Baskin CC. Identification and characterization of the water gap in physically dormant seeds of Geraniaceae, with special reference to *Geranium carolinianum*. Annals of Botany. 2010;**105**:977-990. DOI: 10.1093/aob/mcq078

[11] Todd-Bockarie AH, Duryea ML. Seedpretreatment methods to improve germination of the multipurpose West African forest species *Dialium guineense*. Forest Ecological Management. 1993;**57**:257-273. DOI: 10.1016/0378-1127(93)90176-N

[12] Singh I, Gill MS, Bains TS, Singh P. Genetic analysis of Hardseededness in Mungbean (*Vigna radiata* L. Wilzeck). Crop Improvement-India. 2005;**32**(2):170-172

[13] Marjushkin VF, Sichkar VI, Michailov VG, Polivoda LV. Research notes: Inheritance of hardseededness in soybean. Soybean Genetics Newsletter. 1987;**14**:Article 74, 294-Article307

[14] Smýkal P, Vernoud V, Blair MW, Soukup A, Thompson RD. The role of the testa during development and in establishment of dormancy of the legume seed. Frontiers in Plant Science. 2014;**5**(351):1-19. DOI: 10.3389/fpls.2014.00351

[15] Paulsen TR, Colville L, Kranner I, Daws MI, Högstedt G, Vandvik V, et al. Physical dormancy in seeds: A game of hide and seek? The New Phytologist. 2013;**198**:496-503. DOI: 10.1111/nph.12191

[16] Baskin JM, Baskin CC, Li X. Taxonomy, anatomy and evolution of physical dormancy in seeds. Plant Species Biology. 2008;**15**:139-152. DOI: 10.1046/j.1442-1984.2000.00034.x

[17] Gama-Arachchige NS, Baskin JM, Geneve RL, Baskin CC. Quantitative analysis of the thermal requirements for stepwise physical dormancy-break in seeds of the winter annual *Geranium carolinianum* (Geraniaceae). Annals of Botany. 2013;**111**:849-858. DOI: 10.1093/aob/mct046

[18] Jaganathan GK. Influence of maternal environment in developing different levels of physical dormancy and its ecological significance. Plant Ecology. 2016;**217**:71-79. DOI: 10.1007/s11258-015-0560-y

[19] MacGregor DR, Kendall SL, Florance H, Fedi F, Moore K, Paszkiewicz K, et al. Seed production temperature regulation of primary dormancy occurs through control of seed coat phenylpropanoid metabolism. New Phytologist. 2015;**205**(2):642-652. DOI: 10.1111/nph.13090

[20] El-Keblawy A, Shabana HA, Navarro T, Soliman S. Effect of maturation time on dormancy and germination of *Citrullus colocynthis* (Cucurbitaceae) seeds from the Arabian hyper-arid deserts. BMC Plant Biology. 2017;**17**:1-10. DOI: 10.1186/s12870-017-1209-x

[21] Lachabrouilli AS, Rigal K, Corbineau F, Bailly C. Effects of agroclimatic conditions on sunflower seed dormancy at harvest. European Journal of Agronomy. 2021;**124**:126209. DOI: 10.1016/j.eja.2020.126209

[22] Bentsink L, Koornneef M. Seed dormancy and germination. In: John Herlihy J, Ludwig RN, Ackerveken G, JM MD, editors. The Arabidopsis Book. Vol. 6. Washington US: American Society of Plant Biologists; 2008. pp. 1-18. DOI: 10.1199/tab.0119

[23] Mavi K, Ermiş S, Kenanoğlu BB, Demir İ. Fide yetiştiriciliğinde tohum kalitesi ve tohum uygulamalarının önemi. In: Elliatıoğlu S, Yetişir H, editors. Sebzelerde Fide Yetiştiriciliği. 1st ed. Ankara, Türkiye: Gece Publishing; 2022. pp. 69-134 (in Turkish)

[24] Schopfer P, Plachy C. Control of seed germination by abscisic acid. II. Effect on embryo water uptake in *Brassica napus* L. Plant Physiology. 1984;**766**(1):155-160. DOI: 10.1104/pp.77.3.676

[25] Bewley JD. Seed germination and dormancy. The Plant Cell. 1997;**9**(7):1055. DOI: 10.1105/tpc.9.7.1055

[26] Manz B, Muller K, Kucera B, Volke F, Leubner-Metzger G. Water uptake and distribution in germinating tobacco seeds investigated in vivo by nuclear magnetic resonance imaging. Plant Physiology. 2005;**1386**(1):1538-1551. DOI: 10.1104/pp.105.061663

[27] Nonogaki H, Bassel GW, Bewley JD. Germination-still a mystery. Plant Science. 2010;**179**:574-581. DOI: 10.1016/j.plantsci.2010.02.010

[28] Pereira do Espirito Santo A, Caixeta Oliveira H, Fernandes Fraceto L, Santaella C. Nanotechnology potential in seed priming for sustainable agriculture. Nanomaterials. 2021;**11**:267. DOI: 10.3390/nano11020267

[29] Foschi ML, Juan M, Pascual B, Pascual-Seva N. Influence of seed-covering layers on caper seed germination. Plants. 2023;**12**(3):439. DOI: 10.3390/plants12030439

[30] Debeaujon I, Leon-Kloosterziel KM, Koornneef M. Influence of the testa on seed dormancy, germination, and longevity in Arabidopsis. Plant Physiology. 2000;**122**(2):403-414 10.1104/pp.122.2.403

[31] Weber H, Borisjuk L, Wobus U. Controlling seed development and seed size in *Vicia faba*: A role for seed coat-associated invertases and carbohydrate state. The Plant Journal. 1996;**10**:823-834

[32] Raviv B, Aghajanyan L, Granot G, Makover V, Frenkel O, Gutterman Y, et al. The dead seed coat functions as a long-term storage for active hydrolytic enzymes. PLoS One. 2017;**12**(7):e0181102. DOI: 10.1371/journal.pone.0181102

[33] Oidov A, Sukhee B, Agvaantseren T. Some results of the experiment on scarification of seeds of forage legume (the sample of alfalfa seeds). World Science. 2023;**2**(80):1-8. DOI: 10.31435/rsglobal_ws/30062023/7981

[34] Ermis S, Demir I. Determining the Effect of some Treatments in Relation to Elemination of Hardseededness and Its Maintenance during Storage in Okra, 5. Çanakkale, Turkey: Sebze Tarımı Sempozyumu; 2004. pp. 101-105 (in Turkish)

[35] Avci S, Kaya MD. Seed and germination characteristics of wild Onobrychis taxa in Turkey. Turkish Journal of Agriculture and Forestry. 2013;**37**(5):555-560. DOI: 10.3906/tar-1211-29

[36] Cuadra CDL, Vidal AK, Lagomarsino F, Peñaloza P, Mansur LM, Huenchuleo C. Effect of temperature and scarification on seed germination of *Conanthera* spp. (Tecophilaeaceae). Chilean Journal of Agricultural Research. 2019;**79**(2):323-329. DOI: 10.4067/S0718-58392019000200323

[37] Liyanage GS, Offord CA, Sommerville KD. Techniques for breaking seed dormancy of rainforest species from genus Acronychia. Seed Science and Technology. 2020;**48**(2):159-165. DOI: 10.15258/sst.2020.48.2.03

[38] Górnik K, Sas-Paszt L, Seliga Ł, Pluta S, Derkowska E, Głuszek S, et al. The effect of different stratification and scarification treatments on breaking the dormancy of Saskatoon berry seeds. Agronomy. 2023;**13**(2):520. DOI: 10.3390/agronomy13020520

[39] Keskin B, Temel S, Gurel G, Ozden E. Effects of some temperature and dormancybreaking applications on germination rates of camelthorn (*Alhagi pseudalhagi* (Bieb.) Desv.) seeds. Research in Agricultural Sciences. 2022;**54**(1):22-30. DOI: 10.5152/AUAF.2023.220307

[40] Demir I, Ermis S. Effect of harvest maturity and drying method on okra seed quality. Seed Technology. 2005;**27**:81-88. Available from: https://www.jstor.org/stable/23433218

[41] Okereke CN, Ikegbunam CN, Nwaogaranya UP, Ogbu AC, Francis OA. Iroka CF. Comparative evaluation of the effects of pre-sowing treatments on the germination and growth parameters of *Abelmoschus esculenta* Linn. Asian Journal of Research Crop Science, 2023;**8**(3):108-118. DOI: 10.9734/ajrcs/2023/v8i3172

[42] Paiva PDDO, Silva DPCD, Silva BRD, Sousa IPD, Paiva R, Reis MVD. How scarification, GA_3 and graphene oxide influence the in vitro establishment

and development of Strelitzia. Plants. 2023;**12**(11):2142. DOI: 10.3390/plants12112142

[43] Labbaf N, Rohollahi I, Naji AM. Muscari seed germination enhancement by using sulfuric acid, and stratification priming. Ornamental Horticulture. 2023;**29**:171-180. DOI: 10.1590/2447-536x.v29i2.2548

[44] Pergolotti V, Marcellini M, Contreras E, Mezzetti B, Gambardella M, Capocasa F, et al. Standardization of an in vitro seed germination protocol compared to acid scarification and cold stratification methods for different raspberry genotypes. Horticulturae. 2023;**9**(2):153. DOI: 10.3390/horticulturae9020153

[45] Razanamandranto S, Tigabu M, Neya S, Odén PC. Effects of gut treatment on recovery and germinability of bovine and ovine ingested seeds of four woody species from the Sudanian savanna in West Africa. Flora-Morphology, Distribution, Functional Ecology of Plants. 2004;**199**:389-397. DOI: 10.1078/0367-2530-00167

[46] Jaganathan GK, Yule K, Liu B. On the evolutionary and ecological value of breaking physical dormancy by endozoochory. Perspectives in Plant Ecology, Evolution and Systematics. 2016;**22**:11-22. DOI: 10.1016/j.ppees.2016.07.001

[47] Abbasi M, Heidari M, Rahimi M. Improving germination of guava (*Psidium guajava*) seeds by acid scarification. Journal of Horticultural Science. 2014;**27**(4):394-399. DOI: 10.22067/JHORTS4.V0I0.30581

[48] Gandy YP, Persans MW, Summy KR. An acid-bath technique to break seed dormancy in common sunflower, *Helianthus* L. *annuus* (Asteraceae). Subtropical Agriculture and Environments. 2015;**66**:23-26

[49] Ibrahim HS, Aref Hawramee OK. Impact of acid scarification and cold mist stratification on enhancing seed germination and seedling early growth of *Albizia lebbeck* (L.) Benth. Mesopotamia Journal of Agriculture. 2019;**47**(2):1-13. DOI: 10.33899/magrj.2019.163175

[50] Deltalab B, Moghadam NN, Raad MK, Kaviani B. The effect of cold and acid scarification on seed germination of three green space tree species. Journal of Ornamental Plants. 2023;**13**(2):85-97. Available from: https://dorl.net/dor/28210093.2023.13.2.2.5

[51] Promwee A, Islam SS, Khomphet T. Effect of dissolved oxygen and chemical scarification on *Andrographis paniculata* seed germination in macrobubble conditions. International Journal of Agronomy. 2023;**2023**:1-13. DOI: 10.1155/2023/3459377

[52] Yeh CH, Chen KY, Lee YI. Asymbiotic germination of *Vanilla planifolia* in relation to the timing of seed collection and seed pretreatments. Botanical Studies. 2021;**62**(6):1-12. DOI: DOI.10.1186/s40529-021-00311-y

[53] Ponert J, Figura T, Vosolsobě S, Lipavská H, Vohník M, Jersáková J. Asymbiotic germination of mature seeds and protocorm development of *Pseudorchis albida* (Orchidaceae) are inhibited by nitrates even at extremely low concentrations. Botany. 2013;**91**:662-670. DOI: 10.1139/cjb-2013-0082

[54] Ponert J, Šoch J, Vosolsobě S, Čiháková K, Lipavská H. Integrative study supports the role of trehalose in carbon transfer from fungi to mycotrophic orchid. Front Plant Science. 2021;**12**:793876. DOI: 10.3389/fpls.2021.793876

[55] Kindiger B. A method to enhance germination of eastern gamagrass. Maydica. 1994;**39**:53-53

[56] Hassan SM, Ghareib HR. Bioactivity of *Ulva lactuca* L. acetone extract on germination and growth of lettuce and tomato plants. African Journal of Biotechnology. 2009;**8**(16):3832-3838

[57] Nasreen S, Khan MA, Zia M, Ishaque M, Uddin S, Arshad M, et al. Response of sunflower to various pre-germination techniques for breaking seed dormancy. Pakistan Journal of Botany. 2015;**47**(2):413-416

[58] Agostini RTD, Abrantes FL, Machado-Neto NB, Custódio CC. Ethanol and hormones in physiological conditioning on germination and seed dormancy of Urochloa humidicola cv. Llanero. Journal of Seed Science. 2022;**44**:e202244022. DOI: 10.1590/2317-1545v44261411

[59] Aliloo AA, Darabinejad S. Evaluation of different techniques for breaking seed dormancy of *Heliotropium europaeum* L. (Boraginaceae). Journal of Biological and Environmental Science. 2013;**7**(20):87-91

[60] Rodrigues-Junior AG, Mello ACM, Baskin CC, Baskin JM, Oliveira DM, Garcia QS. A function for the pleurogram in physically dormant seeds. Annals of Botany. 2019;**23**(5):867-876. DOI: 10.1093/aob/mcy222

[61] Baskin CC, Baskin JM. Seeds: Ecology, Biogeography, and, Evolution of Dormancy and Germination. Lexington, Kentucky, US: Academic Press, Elsevier; 1998. 666 p. DOI: 10.1016/B978-0-12-080260-9.X5000-3

[62] Mohammadi G, Khah EM, Honarmand SJ, Shirkhani A, Shabani G. Effects of seed hardness breaking techniques on okra (*Abelmoschus esculentus* L.) germination. International Journal of Agriculture and crop sciences. 2012;**4**(6):264-273

[63] Schelin M, Tigabu M, Eriksson I, Sawadogo L, Oden PC. Effects of scarification, gibberellic acid and dry heat treatments on the germination of *Balanites aegyptiaca* seeds from the Sudanian savanna in Burkina Faso. Seed Science and Technology. 2003;**31**(3):605-617. DOI: 10.15258/sst.2003.31.3.10

[64] Tung LD, Serrano EP. Effects of warm water in breaking dormancy of rice seed. Omonrice. 2011;**18**(1):129-136

[65] Demir I. Development of seed quality during seed development in okra. In: Proceedings of the International Symposium on Agrotechnics and Storage of Vegetable and Ornamental Seeds. Bari, Italy; Vol. 362. Jun 1994. pp. 125-132

[66] Demir I. The effects of heat treatment on hardseededness of serially harvested okra seed lots at optimum and low temperatures. Scientia Horticulturae. 2001;**89**(1):1-7. DOI: 10.1016/S0304-4238(00)00216-8

[67] Giasson C, Baretta C, Sobral LS, Baldissera R. Dormancy breaking, germination, and production of *Mimosa bimucronata* (DC.) Kuntze seedlings. Cerne. 2019;**25**:68-75. DOI: 10.1590/01047760201925012612

[68] Chaodumrikul S, Kaewsorn P, Chulaka P, Chanprasert W. Breaking seed dormancy in smooth loofah (*Luffa cylindrica* (L.) M. Roem.) using scarification and dry heat treatment. Agriculture and Natural Resources. 2016;**50**(2):85-88. DOI: 10.1016/j.anres.2015.09.003

[69] Taghizadeh M, Sajadi FS. Effects of dormancy breaking methods on germination of *Cercis siliquastrum* and

Spartium junceum and seedling growth. Ornamental Horticulture. 2023;**29**:28-36. DOI: 10.1590/2447-536X.v29i1.2528

[70] Baskin JM, Baskin CC. A classification system for seed dormancy. Seed Science Research. 2004;**14**(1):1-16. DOI: /10.1079/ssr2003150

[71] Ellis RH, Hong TD, Roberts EH. Handbook of Seed Technology for Genebanks. Vol. 1. Principles and Methodologies. Rome: IBPGR; 1985. 210 p

[72] Jayasuriya K, Baskin JM, Baskin C. Cycling of sensitivity to physical dormancy-break in seeds of *Ipomoea lacunosa* (Convolvulaceae) and ecological significance. Annals of Botany. 2008;**101**(3):341-352. DOI: 10.1093/aob/mcm285

[73] Li X, Baskin JM, Baskin CC. Anatomy of two mechanisms of breaking physical dormancy by experimental treatments in seeds of two north American Rhus species (Anacardiaceae). American Journal of Botany. 1999;**86**:1505-1511. DOI: 10.2307/2656788

[74] Rodriguez Araujo ME, Pérez D, Aronson J, Cross A. Filling gaps on seed germination and species selection in drylands of Argentina: Work in progress and reflections on intelligent tinkering. Multequina. 2021;**30**:165-180

[75] Nandi OI. Ovule and seed anatomy of Cistaceae and related Malvanae. Plant Systematics and Evolution. 1998;**209**:239-264. Available from: https://www.jstor.org/stable/23643179

[76] Brits GJ, Manning JC. Seed structure and physiology in relation to recruitment ecology in Leucospermum (Proteaceae) in fynbos. Australian Journal of Botany. 2019;**67**(4):290-308. DOI: 10.1071/BT18199

[77] Van Staden J, Jäger AK, Light ME, Burger BV, Brown NAC, Thomas TH. Isolation of the major germination cue from plant-derived smoke. South African Journal of Botany. 2004;**70**(4):654-659. DOI: 10.1016/S0254-6299(15)30206-4

[78] Kulkarni MG, Light ME, Van Staden J. Plant-derived smoke: Old technology with possibilities for economic applications in agriculture and horticulture. South African Journal of Botany. 2011;**77**(4):972-979. DOI: 10.1016/j.sajb.2011.08.006

[79] Khatoon A, Rehman SU, Aslam MM, Jamil M, Komatsu S. Plant-derived smoke affects biochemical mechanism on plant growth and seed germination. International Journal of Molecular Sciences. 2020;**21**(20):7760. DOI: 10.3390/ijms21207760

[80] Gokdas Z, Yildirim E, Gupta S, Demir I. Karrikinolide stimulated seed germination of artificially aged marrow, cabbage and pepper seeds through repair of cell structure and enzyme activity. South African Journal of Botany. 2022;**151**:208-213. DOI: 10.1016/j.sajb.2022.09.049

[81] Hradilová I, Duchoslav M, Brus J, Pechanec V, Hýbl M, Kopecký P, et al. Variation in wild pea (*Pisum sativum* subsp. *elatius*) seed dormancy and its relationship to the environment and seed coat traits. PeerJ. 2019;**7**(e6263):2-32. DOI: 10.7717/peerj.6263

[82] Tavşanoğlu Ç, Ergan G, Çatav ŞS, Zare G, Küçükakyüz K, Özüdoğru B. Multiple fire-related cues stimulate germination in *Chaenorhinum rubrifolium* (Plantaginaceae), a rare annual in the Mediterranean Basin. Seed Science Research. 2017;**27**(1):26-38. DOI: 10.1017/S0960258516000283

[83] Downes KS, Light ME, Pošta M, Van Staden J. Fire-related cues and

the germination of eight Conostylis (Haemodoraceae) taxa, when freshly collected, after burial and after laboratory storage. Seed Science Research. 2015;**25**(3):286-298. DOI: 10.1017/S0960258515000227

[84] Liyanage GS, Ooi MK. Do dormancy-breaking temperature thresholds change as seeds age in the soil seed bank? Seed Science Research. 2017;**27**(1):1-11. DOI: 10.1017/S0960258516000271

[85] Harrington GT. Agricultural value of impermeable seeds. Journal of Agricultural Research. 1916;**6**:761-796

[86] Roh MS, Bentz JA, Wang P, Li E, Koshioka M. Maturity and temperature stratification affect the germination of Styrax japonicus seeds. The Journal of Horticultural Science and Biotechnology. 2004;**79**:645-651. DOI: 10.1080/14620316.2004.11511820

[87] Ozden E, Light ME, Demir I. Alternating temperatures increase germination and emergence in relation to endogenous hormones and enzyme activities in aubergine seeds. South African Journal of Botany. 2021;**139**:130-139. DOI: 10.1016/j.sajb.2021.02.015

[88] Fenner M, Thompson K. The Ecology of Seeds. Cambridge, UK: Cambridge University Press; 2005. 250 p. DOI: 10.1017/CBO9780511614101

[89] Widell KO, Vogelmann TC. Fiber optic studies of light gradients and spectral regime within *Lactuca sativa* achenes. Physiologia Plantarum. 1988;**72**(4):706-712. DOI: 10.1111/j.1399-3054.1988.tb06369.x

[90] Brady SM, McCourt P. Hormone cross-talk in seed dormancy. Journal of Plant Growth Regulation. 2003;**22**:25-31. DOI: 10.1007/s00344-003-0018-7

[91] Casal JJ, Sánchez RA. Phytochromes and seed germination. Seed Science Research. 1998;**8**(3):317-329. DOI: 10.1017/S0960258500004256

[92] Sanchez RA, Mella RA. The Exit from Dormancy and the Induction of Germination: Physiological and Molecular Aspects. Handbook of Seed Physiology Application to Agriculture. 1st ed. New York: Food Product Press and The Haworth Press, Inc; 2004. 479 p

[93] Lamont BB, Pausas JG. Seed dormancy revisited: Dormancy-release pathways and environmental interactions. Functional Ecology. 2023;**37**(4):1106-1125. DOI: 10.1111/1365-2435.14269

[94] Motsa MM, Slabbert MM, Van Averbeke W, Morey L. Effect of light and temperature on seed germination of selected African leafy vegetables. South African Journal of Botany. 2015;**99**:29-35. DOI: 10.1016/j.sajb.2015.03.185

[95] Gao S, Chu C. Gibberellin metabolism and signaling: Targets for improving agronomic performance of crops. Plant and Cell Physiology. 2020;**61**(11):1902-1911. DOI: 10.1093/pcp/pcaa104

[96] Kaneko M, Itoh H, Ueguchi-Tanaka M, Ashikari M, Matsuoka M. The α-amylase induction in endosperm during rice seed germination is caused by gibberellin synthesized in epithelium. Plant Physiology. 2002;**128**(4):1264-1270. DOI: 10.1104/pp.010785

[97] Shaik SS, Carciofi M, Martens HJ, Hebelstrup KH, Blennow A. Starch bioengineering affects cereal grain germination and seedling establishment. Journal of Experimental Botany. 2014;**65**(9):2257-2270. DOI: 10.1093/jxb/eru107

[98] Thomas TH. Some reflections on the relationship between endogenous hormones and light-mediated seed dormancy. Plant Growth Regulation. 1992;**11**:239-248

[99] Lee SY, Park K, Jang BK, Ji B, Lee H, Baskin CC, et al. Exogenous gibberellin can effectively and rapidly break intermediate physiological dormancy of *Amsonia elliptica* seeds. Frontiers in Plant Science. 2022;**13**:1043897. DOI: 10.3389/fpls.2022.1043897

[100] Ozden E, Demir I. GA_3 enhanced seed germination of *Solanum torvum*. Radovi Poljoprivrednog Fakulteta Univerziteta u Sarajevu (Works of the Faculty of Agriculture University of Sarajevo. 2016;61:316-320

[101] He XQ, Wang YR, Hu XW, Baskin CC, Baskin JM, Lv YY. Seed dormancy and dormancy-breaking methods in *Leymus chinensis* (Trin.) Tzvel (Poaceae) Grass and Forage. Science. 2016;**71**(4):641-648. DOI: 10.1111/gfs.12220

[102] Chen JZ, Huang XL, Xiao XF, Liu JM, Liao XF, Sun QW, et al. Seed dormancy release and germination requirements of *Cinnamomum migao*, an endangered and rare woody plant in Southwest China. Frontiers in Plant Science. 2022;**13**:770940. DOI: 10.3389/fpls.2022.770940

[103] Silva ECD, Villa F, Silva DFD, Possenti JC, Silva LSD, Ritter G. Araticum accessions: Effect of gibberellic acid concentrations and soaking times on seed dormancy overcoming. Revista Caatinga. 2021;**34**:614-620. DOI: 10.1590/1983-21252021v34n313rc